Make: DIY Drone and Quadcopter Projects

A Collection of Drone-Based Essays, Tutorials, and Projects

From the Editors of Make:

Make: DIY Drone and Quadcopter Projects

by The Editors of Make:

Printed in the United States of America.

Published by Maker Media, Inc., 1160 Battery Street East, Suite 125, San Francisco, CA 94111.

Maker Media books may be purchased for educational, business, or sales promotional use. Online editions are also available for most titles (*http://safaribooksonline.com*). For more information, contact O'Reilly Media's institutional sales department: 800-998-9938 or *corporate@oreilly.com*.

Editor: Patrick Di Justo
Production Editor: Colleen Cole
Proofreader: Charles Roumeliotis
Indexer: WordCo Indexing Services
Interior Designer: David Futato
Cover Designer: Ryland Grudzinski
Illustrator: Rebecca Demarest

April 2016: First Edition

Revision History for the First Edition

2016-04-05: First Release

See *http://oreilly.com/catalog/errata.csp?isbn=9781680451290* for release details.

978-1-680-45129-0

[LSI]

Table of Contents

PART I. Features

PART I

Features

The first section of the book is a collection of feature articles from the pages of *Make:*, providing an introduction to the world of drones: those autonomous aerial vehicles that mostly (but not always) take the form of small quadrotor helicopters called quadcopters.

The first article looks into what exactly drones are, and more importantly, what they are not. The second article provides a detailed breakdown of the anatomy of a drone: from its airframe to its propellers, from its battery to its brain. Subsequent articles look at how drones work, the positive ways drones are being used —as a way to drop food and other supplies to refugees in war-torn places around the world, and other uses for remote flying vehicles. With these five articles under your belt, you'll be in a great position to understand and enjoy the rest of the book!

Mind Your Drone

1

Not everybody likes the word drone—industry and military experts avoid using the "D word" in public. They prefer the term "unmanned aerial system or vehicle" and associated acronyms UAS or UAV. Despite plenty of misgivings about military and spy drones, the word drone has become widespread and popular, used with great enthusiasm by hobbyists who hang out on the DIY Drones site and by professional aerial photographers like the L.A.-based Drone Dudes. So what is a drone?

—Dale Dougherty, founder of Make: Magazine and creator of Maker Faire

The original meaning of drone is a male bee. The body of a drone is bigger than all other bees (except the queen), but what physically distinguishes a drone is a larger pair of compound eyes. Yet drones have no real work to do but reproduce. They make late afternoon flights to what is called a congregation area, where drones gather looking to mate with a virgin queen. Once these bees succeed, however (and they perform this act in midair), the drones fall out of the sky, having left an essential body part behind. That's all that drones do.

The notion that a drone doesn't have much work of its own leads to a secondary definition of a drone as someone who lives off the work of others—a parasite. In fact, at the end of summer, the worker bees kick the remaining drones out of the hive. They eat too much and do too little. They can be replaced in the spring.

This helps set up the problem. We not only need to figure out a definition for drones, we also have to figure out what they're going to do—and not do. While some agree that drones are unmanned, others point out that they're piloted, preferring the acronym RPA for "remotely piloted aircraft." That wouldn't differentiate drones from remote-control aircraft, but it emphasizes that a human, who can be held responsible, is at the controls. A drone can be operated manually or it can be programmed to follow a fixed flight plan.

The distinguishing feature of a drone seems to be the promise of autonomy. Today, a typical flight consists of switching between manual flight and autopilot. How much further might it go? Given the right instrumentation and the ability to process that data, could a drone be programmed to make context-aware decisions, particularly ones that humans are not very good at? A drone might detect problems before they occur, such as responding to gusts of wind or avoiding unexpected obstacles. A drone might also be able to communicate with other drones.

Can a drone be considered a robot, able to obey Asimov's Three Laws of Robotics? We need drones that explicitly avoid harming humans and can act to protect themselves from destruction. We should expect this much from any fully autonomous vehicle. A drone then might be said to have a mind of its own.

Until such time, however, that responsibility falls on the person flying the drone. When you fly a drone, you aren't just a user—you're a pilot. You must protect your equipment, yourself, and most importantly, other people. A bad or incompetent pilot can injure people or invade their privacy. It's not a lot different from owning a pet or a car.

Good pilots, like the Drone Dudes, worry that bad or careless pilots will garner the public's attention, create a climate of fear, and cause governments to restrict or eliminate drones for commercial or recreational use. The reason we need better technology is that few of us are very good pilots.

For makers, the most interesting challenge isn't just building drones or flying them. It's discovering what drones are good for, what creative uses they have, and what tough problems they might solve. Otherwise, planes and quadcopters will be sold only as toys, not tools, and many people will discard them once they lose interest in their playthings. We're hoping drones become platforms for developing compelling applications that will push the technology forward and adjust the balance between the light and the dark side of drones.

Getting Started with Multicopters

2

Tips on how to build, buy, fly, and spy with multirotor R/C helicopters.

—From Make:31 by Frits Lyneborg

A multicopter is a flying robot resembling a wagon wheel—without the wheel. It has a central hub with electronics, power, and sensors, onto which are mounted arms that hold propellers to provide lift. The number of arms gives the name: a tricopter (trirotor) has three arms, a quadrocopter or quadcopter (quadrotor) has four, a hexacopter six, and an octocopter eight. There are other variations, but these are the most popular setups.

They're also called multirotors, which arguably is the correct term, but I'll stick to multicopters because that's used more often on the Internet, where you'll find the most information on the topic. Why try multicopters? Perhaps you saw one and you just have to own this cool new gadget. Or you fly R/C planes and you'd like to try a new type of aircraft. Or you're into DIY electronics or robots, or you want to do aerial photography. Whatever your motivation, there's an option for you. I've flown a variety of multicopters and built three of my own, so I've picked up a few tips I can share.

Homebrew Pedigree

In 2003, Hong Kong-based company Silverlit Electronics read in the newspaper about students Daniel Gurdan and Klaus M. Doth's prize-winning entry in Germany's national Young Scientists competition. Gurdan and Doth's design was of a radio-controlled, self-leveling quadrocopter (Figure 2-1).

Figure 2-1 *Gurdan and Doth*

In late 2004, Silverlit began production of their X-UFO, a simplified and cheaper version of the students' design (Figure 2-2). When this product hit the international markets over the next few years, it seeded the idea of a small, remote controlled multicopter to many people throughout the world. Today there are dozens on the market.

Figure 2-2 *The Silverlit X-UFO*

How They Work

On an ordinary helicopter, the tail rotor provides horizontal thrust to counteract the main rotor's torque, in order to keep the helicopter from spinning around with the main blades (Figure 2-3).

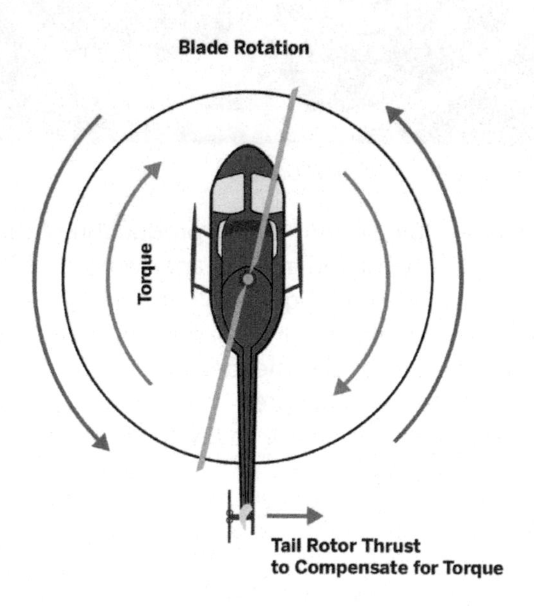

Figure 2-3 *How a regular helicopter works*

A multicopter works quite differently. Take for instance a quadrocopter: every second propeller spins in the opposite direction (Figure 2-4), counteracting the torque of the adjacent propellers.

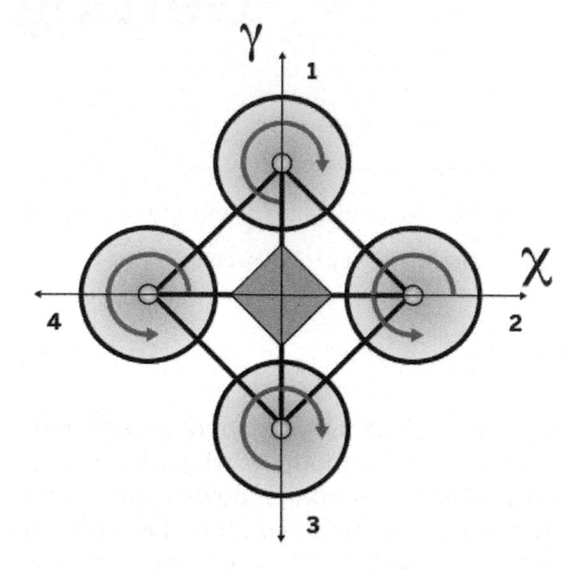

Figure 2-4 *Quadcopter propellers spin in opposite directions*

More importantly, a multicopter has an onboard computer that varies the speed on individual propellers, making possible every form of spin, tilt, yaw, and rudder control around any center and any axis, as well as flight in any direction.

Your First Multicopter

The best starter multicopter is lightweight: the lighter the copter, the less damage to it and to the surroundings when you crash. And you will crash! The bigger they get, the more scared you'll be of flying them. Large multicopters can rip through clothes and flesh, and they cost a lot of money. The downside is that lighter versions carry less payload (read: cameras and extra sensors), and flying time is usually shorter. The upside is that they're cheaper. Ironically, it's also a good thing that lighter copters are typically harder to control, due to fewer sensors and less-sophisticated overall construction. Why is this good? Because you'll learn to fly. A heavy, complex autonomous multicopter might be easier to fly—or even fly by itself—but you'll never learn to handle a multicopter that way. That can be a big problem the moment something goes wrong. And something will go wrong.

Building Your Own Multicopter

Once you've played with multicopters, you'll realize that building one is a project that you could take on. Here are the basics.

Batteries and Motors

The real magic here is the combination of the very powerful lithium polymer (LiPo) batteries and brushless motors. These two components, with just a normal R/C plane propeller on the motor, can lift themselves right off the ground, and so this combination can make virtually anything fly.

ESCs and Control Board

A multicopter's flight must be controlled and balanced in a certain way. The motors are controlled by little units called electronic speed controllers (ESCs), and these need signals telling them how much power to pass on. In a multicopter, that signal comes from a special control board. The control board is hooked up to a standard R/C plane receiver, and possibly other peripherals such as GPS, or whatever your imagination and wallet allow. Probably the two most popular control boards right now are HobbyKing's Multi-Rotor Control Board V2.1 (hobbyking.com) and Multi RC Shop's KK Plus V5.5e Multicopter Controller (multircshop.com), both based on Atmel's ATmega168 microcontroller chip (Figure 2-5). Arduino-oriented makers might prefer DIY Drones' ArduCopter system (*http://copter.ardupilot.com/*), with its ArduPilot Mega board based on the ATmega2560 (Figure 2-6).

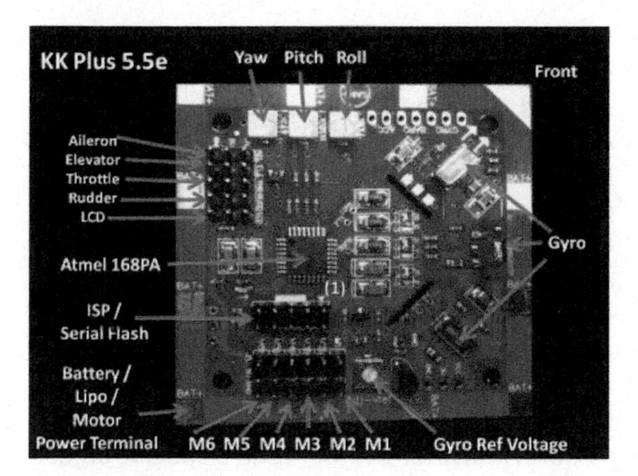

Figure 2-5 *The KK Plus controller*

Figure 2-6 *The ArduCopter*

Body and R/C Gear

The body of a multicopter can be made of almost anything, including wood, so the only "mysterious" thing is the control board. The rest is common R/C gear: a four-channel transmitter and receiver, and connectors to hook up your components. A Google search on "multicopter control board" will get you started and lead you to plenty of build instructions, and I recommend visiting *http://rcexplorer.se*, *http://hobbyking.com*, and *http://diydrones.com*.

Video from a Multicopter

Filming from the sky is the most common broken dream among multicopter users. Unfortunately a lot of people are spending a lot of money hoping to make great professional video from the air at a fraction of the cost of a real helicopter. Many shops out there are ready to sell this dream, which I think is unfair. You should think twice. Here's a test: take your camera and put it on a broomstick. Hold the other end of the broomstick. Now try to get good footage out of that. While it may give interesting new angles and be "arty," in general it's going to look "filmed from the end of a broomstick." You'll find it hard to get the quality of shots you're used to. The same is the case with a multicopter. You can find cool-looking videos made from multicopters on YouTube, but they're always focused on the flight experience ("Look, I'm flying!"), rather than a specific object or person being photographed. If you work hard with your equipment, you can get cool shots, but they'll be lucky shots, unless your copter can transmit video back to the ground (see the section "Cameras and Video Downlinking"). If you get a picture of a house, it'll be awkwardly framed. If you video anything other than random treetops, the subject won't be well placed in the frame, and everything will be moving about. It's not easy.

Gimbals and Gyros

You can purchase very expensive camera mounts and gimbals with gyroscopic stabilization. But before you do, ask to see raw film of at least one minute made with the equipment—not filmed at high speed and slowed down for a smooth look, and not edited in short clips, or stabilized in post-production. I don't recommend two-axis gyro gimbals. In my experience they introduce more shaking than they do good, even the very expensive ones. (And three-axis gimbals introduce even more.) Since multicopters are extremely steady when it comes to holding direction, I don't think these are of any benefit. Your best mount is something simple like a flexible plastic tube or soft foam. Just accept that the camera is not level at all times.

Cameras and Video Downlinking

You can get really cool videos and pictures from multicopters if you've practiced flying, and if you use the medium on its own terms: accept the ever-moving picture, use a lightweight camera, and focus on action shots where the camera is moving through the air. The best videos I've seen use extreme wide-angle shots, usually made with the GoPro camera brand (Figure 2-7), which can also shoot at 60 frames per second (fps), giving a slow-motion feeling. The lighter the camera, the better the flight performance. Think eight ounces and below. Finally, your best tool is video feedback. Actually seeing what you film, while you're doing it, is called first-person video (FPV). There are many options for wireless video downlinks, depending on the following parameters.

Figure 2-7 *A GoPro mounting rig for a drone*

Cost, weight, and power consumption

How large an antenna can you carry to the field? What RF bands are allowed in your country? Which are already used on your copter?

Transmitting power

Systems one watt or stronger may require a ham operator's license. Frequency regulation information is available at *http://makezine.com/go/hamradio*.

Electromagnetic pulses

Powerful transmitters can make servos and other electronics malfunction. These things have to be experienced; there are no golden rules that I'm aware of. Sometimes things just interfere.

In general you're looking for lighter weight, longer range, less power consumption, and undisturbed frequencies. You can't expect to use cheap, random TV transmission gear. Get something from a shop that has experience with video downlinks from multicopters. And if you get a pair of video glasses for monitoring (Figure 2-8), you can see what the camera in the sky sees, even in sunlight. If nothing else, it's really cool to be able to elevate your field of vision by remote control.

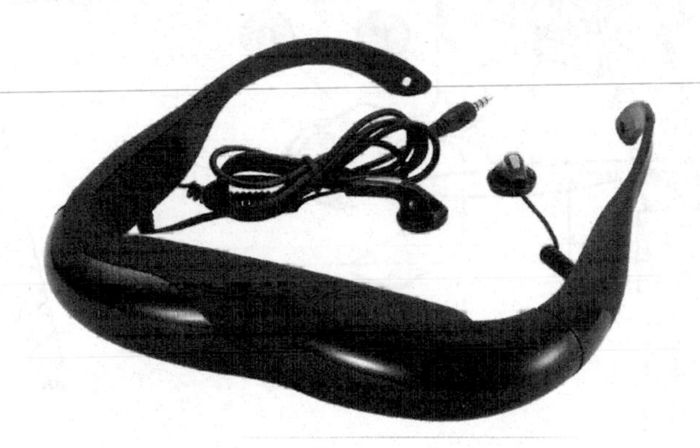

Figure 2-8 *Video glasses for first-person video*

Going Further: Drone Multicopters

Once you've mastered R/C multicopters, you might want to try drone multicopters. When most people say drone they're talking about flying by GPS coordinates and waypoints in fully autonomous mode, and that's something special. One example is the ArduCopter, controlled by an Arduino-based autopilot developed by DIY Drones (Figure 2-9).

Figure 2-9 *Some ArduCopter screens*

There are also popular setups where cameras film the drone, and a computer calculates its flight from what the cameras see (little dots on the copter). Perhaps you could even set up a drone to navigate by the sun. It's all just sensors. If you do experiment with drones, never let your autonomous machine go beyond visual contact. Most systems I know of have a built-in maximum range of 250 meters. Once you start playing with multicopters, you'll notice there's no longer a sharp border between "autonomous" or "R/C" flight. Any multicopter is a robot that to some degree is autonomously controlling its motors (or it would crash). And even fully autonomous drones have the option of killing the automation and returning to R/C control (anything else would be hazardous). With multicopters, it's always some form of R/C, and it's always some kind of autonomous.

More DIY multicopters and kits: *http://scoutuav.com, http://multiwiicopter.com*

Quadcopter FPV: *http://makezine.com/go/fpv*

3D-printable quad: *http://makezine.com/go/hugin*

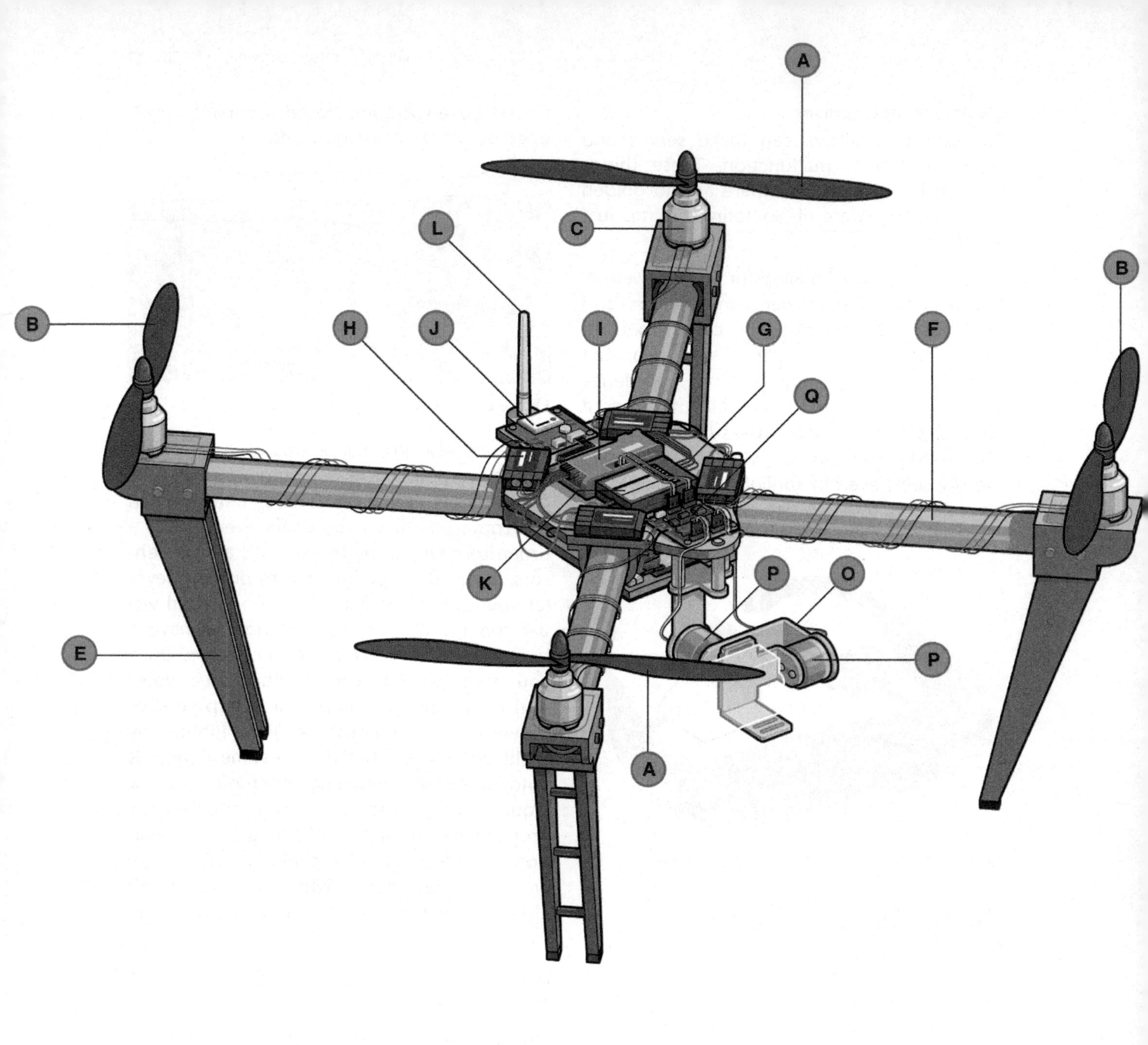
A
C
L
B
H
J
I
G
F
B
Q
K
P
O
E
P
A

Anatomy of a Drone

3

—*Multirotor UAV diagram by Rob Nance*

Standard Prop

The same "tractor" propeller used on standard front-engine R/C airplanes. In orange in the diagram above.

"Pusher" Prop

These contra-rotating props exactly cancel out motor torques during stationary level flight. Opposite pitch gives downdraft. In dark grey in the diagram above.

Motor

Usually a brushless electric "outrunner" type, which is more efficient, more reliable, and quieter than a brushed motor (Figure 3-1).

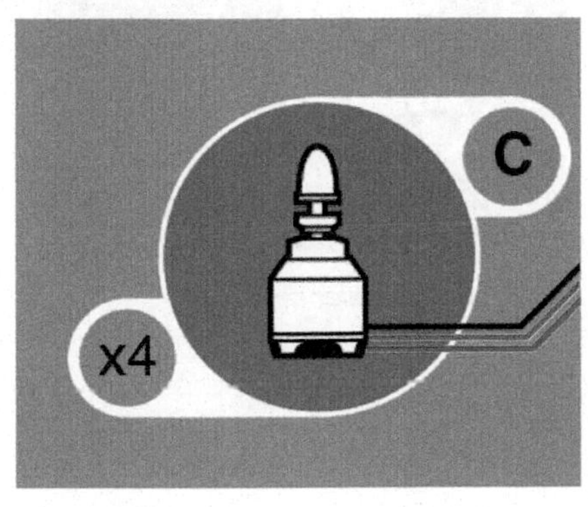

Figure 3-1 *Motor*

Motor Mount

Sometimes built into combination fittings with landing struts (Figure 3-2).

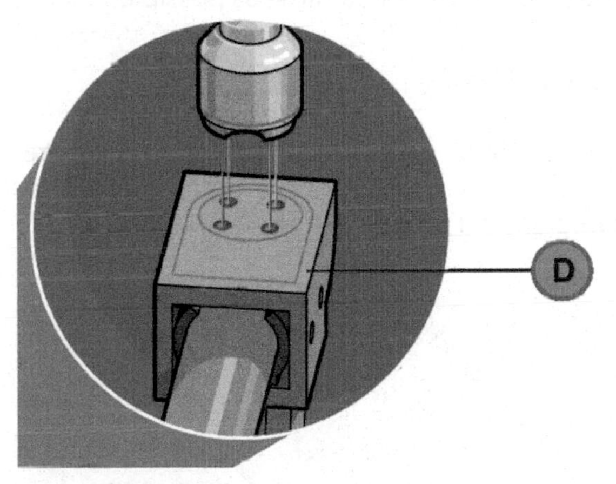

Figure 3-2 *Close-up of motor mount*

Landing Gear and Boom

Designs that need high ground clearance may adopt helicopter-style skids mounted directly to the body, while designs with no hanging payload may omit landing gear altogether (Figure 3-3).

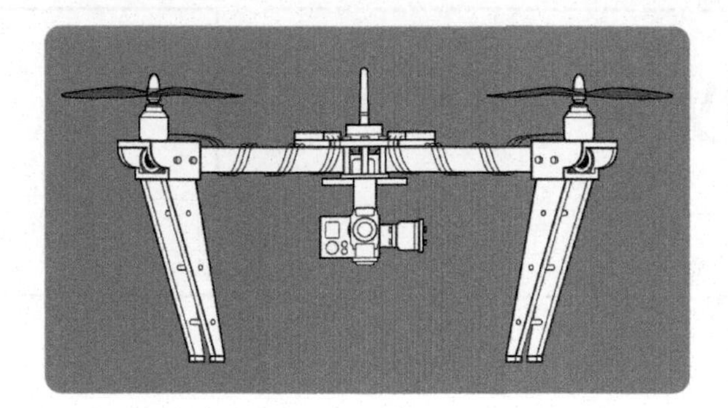

Figure 3-3 *Landing gear and boom*

Boom

Shorter booms increase maneuverability, while longer booms increase stability. Booms must be tough to hold up in a crash while interfering with prop downdraft as little as possible.

Main Body

Central "hub" from which booms radiate like spokes on a wheel. Houses battery, avionics, cameras, and sensors (Figure 3-4).

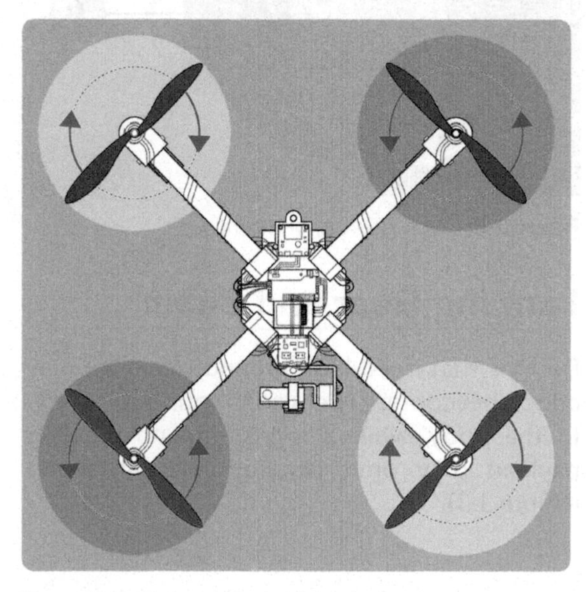

Figure 3-4 *The main body of the drone*

Electronic Speed Controller (ESC)

Converts DC battery power into three-phase AC for driving brushless motors (Figure 3-5).

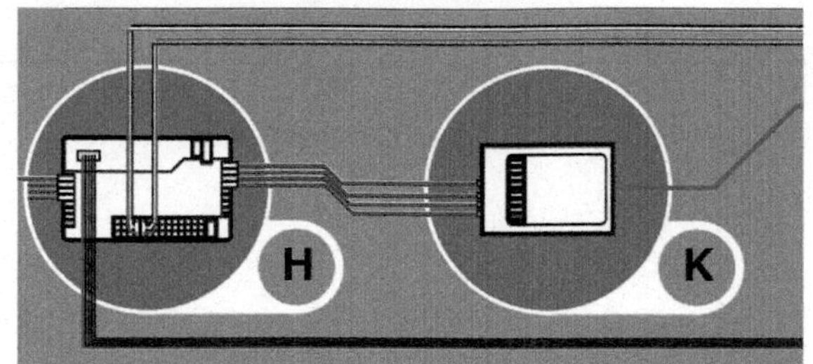

Figure 3-5 *Electronic speed controller (H) and radio receiver (K)*

Flight Controller

Interprets input from receiver, GPS module, battery monitor, and onboard sensors. Regulates motor speeds, via ESCs, to provide steering, as well as triggering cameras or other payloads. Controls autopilot and other autonomous functions (Figure 3-6).

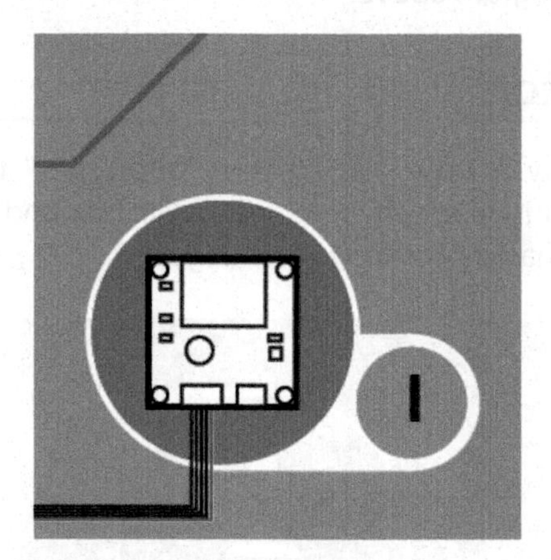

Figure 3-6 *Flight controller computer*

GPS Module

Often combines GPS receiver and magnetometer to provide latitude, longitude, elevation, and compass heading from a single device (Figure 3-7).

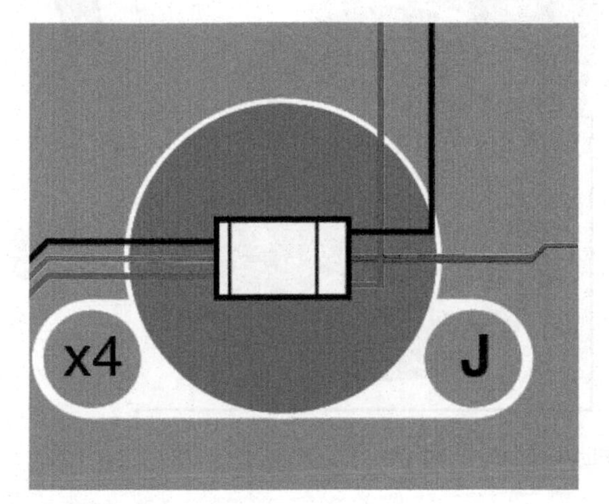

Figure 3-7 *GPS module*

Receiver

Often a standard R/C radio receiver unit. The minimum number of channels needed to control a quad is four, but five is usually recommended (as seen in Figure 3-5).

Antenna

Depending on your receiver, may be a loose wire whip or helical "rubber ducky" type (Figure 3-8).

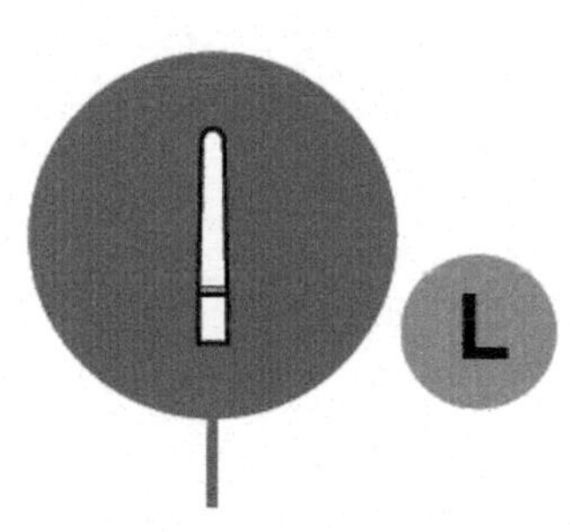

Figure 3-8 *Antenna*

Battery

Lithium polymer (LiPo) batteries offer the best combination of energy density, power density, and lifetime on the market (Figure 3-9).

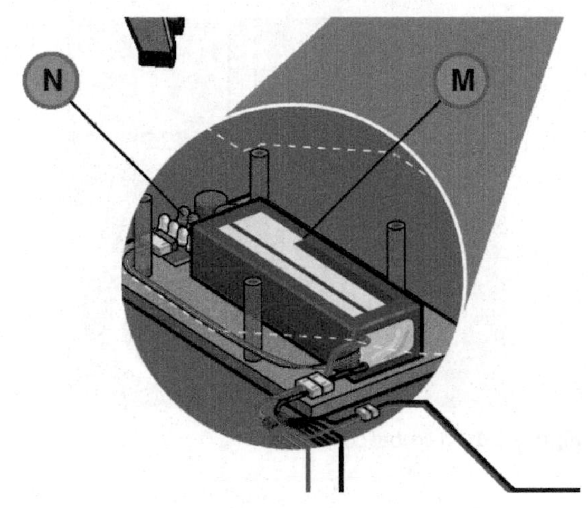

Figure 3-9 *Battery (M) and battery monitor (N)*

Battery Monitor

Provides in-flight power level monitoring to flight controller.

Gimbal

Pivoting mount that rotates about one, two, or three axes to provide stabilization and pointing of cameras or other sensors.

Gimbal Motor

Brushless DC motors can be used for direct-drive angular positioning, too, which requires specially wound coils and dedicated control circuitry that have only recently become commercially available.

Gimbal Controller

Allows control of direct-drive brushless gimbal motors as if they were standard hobby servos (Figure 3-10).

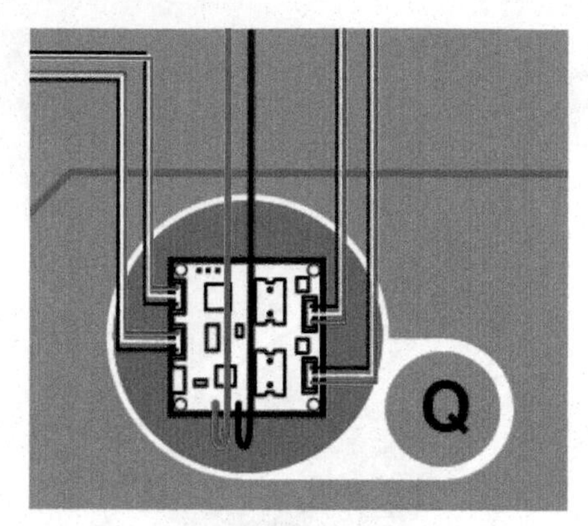

Figure 3-10 *Gimbal controller*

Camera

GoPro or other compact HD video unit with onboard storage. Real-time streaming is possible with special equipment (Figure 3-11).

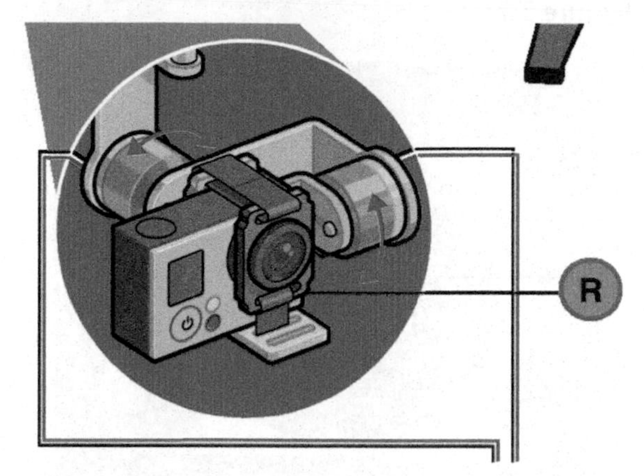

Figure 3-11 *Camera*

Building Drones to Deliver Medicine and Food to War-Torn Syria

4

Fixed-wing UAVs aren't just for bombs—meet the nonprofit building drones that fly supplies to Syria.

—From Make:47 by Signe Brewster

On March 16, 2015, barrels of chlorine gas rained down on the town of Sarmin in northern Syria, killing six and wounding many more—just one of many horrific chemical attacks in the civil war that has consumed the country.

"Sarmin isn't far from the border but the border is closed to all traffic," Sasha Ghosh-Siminoff, president of Syrian aid organization People Demand Change, texted his Stanford University-based friend Mark Jacobsen, four hours after the attack. "If your planes were ready, you could have flown in emergency medicine and gear."

The planes Ghosh-Siminoff was referring to are drones, built expressly for this purpose. Jacobsen is the executive director of Uplift Aeronautics, a nonprofit that hopes to deliver essential medical supplies, food, and other cargo to Syrians via its Syria Airlift Project. Syria recently closed its border to foreign aid, and any planes that attempt to fly over the country run a high chance of being shot down. Uplift has a different plan: fleets of drones that could swoop in by night, undetected by human eyes or radar.

Doing More

Jacobsen, who is pursuing a Ph.D. in political science, was in Istanbul about a year ago with a group of academics when a heated discussion broke out about international intervention in the Syrian war. Since 2011's Arab Spring, when activists came together to protest president Bashar al-Assad and his government, at least 200,000 people have died there. More than 10,000 were children. A lack of medical care and food are among the government's weapons against its own people.

Person after person at the gathering asked the same question: why isn't more being done? Jacobsen, a former Air Force cargo pilot, explained to one attendee that you simply can't fly a cargo plane into such an unpredictable place. It's impossible.

He went back to his hotel that night feeling guilty. It didn't seem like a good enough answer. While speaking with his colleagues, he became fixated on the idea of sending in large numbers of packages—perhaps via drone. He took out a notebook at around 2 or 3 a.m., the hope of sleep long forgotten.

"It seemed like I was onto something with the idea of swarming small packets, but I didn't really know what technology could do that, whether it would be quadcopters or planes or catapults or anything else. Balloons?" Jacobsen says. "I was just trying to lay out everything I could think of."

Uplift Aeronautics and the Syria Airlift Project were born, and today Jacobsen and a group of volunteers are busy flying prototype drones. Their plan is to fly over the border from a neighboring country, on missions chosen by aid partners such as People Demand Change. Each drone can carry only a few pounds of supplies, but their small size makes them untrackable by radar and dispensable. If a chlorine bomb explodes, medicine-carrying drones can be there in an hour, as opposed to days—or never.

Uplift plans to train Syrian refugees and other people on the ground to fly and repair the drones. Its first destination would be Aleppo, Syria's largest city. The war has hit it hard. Hunger and disease are common.

The drones would take about a half hour to fly to Aleppo. Instead of touching down, they would drop their cargo in a small box attached to a parachute. Then they would return. Back at the launch base, the location of which would likely shift from day to day, volunteers could switch out their battery, load new cargo, and launch again within minutes.

Complications

Flying anything, let alone hundreds of drones, into a country without permission is a breach of international law. Current sanctions bar sending US goods into the country. In extreme times like these, exceptions can be granted, but they depend on various government channels.

Jacobsen isn't exactly sure how Uplift will secure an OK from the US, though he has initiated conversations with officials. The drones will likely have to be approved by the US Treasury and international agreements, and will need to comply with arms regulations and counterterrorism laws.

Uplift will also need to talk with the governments in countries bordering Syria, such as Turkey or Jordan. They will need to prove that the drones will be safe and beneficial. The recent election in Turkey, and the country's air strikes within Syria, add a new layer of complexity.

Farm Drones Take Flight

"In some ways, negotiating with the armed groups and the people inside Syria is easier than the Turkish governments," Ghosh-Siminoff says. "It's really difficult to navigate that bureaucracy and know you're in the clear and not running afoul of some archaic rule."

In Syria, the groups fighting Assad would be most likely to shoot down a drone. Currently, the resistance occupies the ground between Uplift's launch site and Aleppo. If Uplift can demonstrate the planes are for aid, and will not interfere with the opposition's efforts, Ghosh-Siminoff said there should not be a problem convincing the locals to let them pass.

In a country strapped for resources, a scenario could arise where troops start capturing drones to use for their own purposes. Uplift thought of that. The drones are equipped with a self-destruct device designed to fry their navigation system if they fly too close to the ground anywhere but at the launch site. Any drone that gets too low will never be remotely pilotable again.

"We're not planning to talk to them at all once they leave the takeoff area," says Jacobsen. "Routes will be pre-programmed. Our custom firmware on the plane actually plugs its ears and stops listening to incoming messages while in Syrian airspace, which should make it considerably harder to hack."

Figure 4-1 *Michael Taylor, with Uplift Aeronautics' Waliid drone mounted on a home-built PVC launcher*

Figure 4-2 *A volunteer tests the drone radio controls*

A Team of Volunteers

On a hot, cloudless day in April, Jacobsen and four volunteers gathered at Stanford University's Lake Lagunita. Engineer Michael Taylor, a Ph.D. candidate in electrical engineering, led two other volunteers through setting up the drone launcher on the lake bed, which has become a grassy field thanks to California's drought.

On a porch overlooking the lake, Jacobsen assembled and tested the "Waliid" drone. He ran new volunteer Stuart Ginn, a medical resident still clad in scrubs, through the plane's software and pre-flight protocol.

Made of foam and held together by tape, the drone is not visually impressive. It's shaped like a plane, as opposed to the quadcopters that have taken over the consumer market, which allows it to fly for an hour instead of minutes. Its wingspan measures 5′ 7″ and is decorated in black, green, and red—the colors of the Syrian flag.

Back in the field, Taylor and aeronautics and astronautics Ph.D. student Heather Kline had completed the launcher—a 7-foot-long PVC pipe skeleton that guides the drone into the air. Tomoki Eto, a mechanical engineering undergraduate and experienced drone pilot, anchors a bungee line to the ground several hundred feet away, stretches it to the launcher, and attaches it. Upon release, the bungee will fling the UAV into the air.

The team consists of five volunteers, but expands to 15 or 50 people, depending on how you look at it. The engineering core resides at Stanford, but people all over the world are contributing to its design and deployment. It's been an informal collaboration via email, Skype, and Dropbox, but Uplift plans to release as much open source material as possible, probably via Github.

Like many of the volunteers, lead engineer Brandon Fetroe got involved with the project after hearing about it through Stanford's UAV club. While the technical hurdles felt manageable to him as an engineer, he says, its political challenges were things "many Americans didn't feel capable of tackling on our own."

Figure 4-3 *Uplift Aeronautics members Heather Kline, Tomoki Eto, Mark Jacobsen, and Michael Taylor*

"The project made it clear from the start that each individual person who was interested in helping out has the opportunity to contribute in ways that match their skill set and that together, as a whole, the team could do things that the individual on their own certainly can't," Fetroe says.

Fetroe, a mechanical engineering Ph.D. student, has been flying R/C planes since he was 12. He described his expertise as a little bit of everything—something that holds true throughout the Uplift team. Ginn, for example, was once a commercial pilot; he's now helping reach out to medical NGOs. And Jacobsen is leveraging his international contacts and friends in the US government from his days in the Air Force.

Breaking New Ground

Interest in using drones for deliveries is high around the world. Syria is just one of many regions where broken infrastructure can make supplies impossible to deliver by land. Drones are already busy monitoring poachers and providing aerial intelligence in disaster situations.

But Jacobsen didn't relate the Syria Airlift Project to any of those efforts. Instead, he looked back much further, to the Cold War when the Western Allies airlifted supplies into West Berlin. US Air Force pilot Gail Halvorsen started a movement when he began dropping candy attached to handkerchief parachutes for children. Like the Candy Bomber, as Halvorsen became known, the drones could drop symbols of hope and happiness.

"People inside Syria affiliate airplanes with death. There are no positive memories of an airplane anymore," Ghosh-Siminoff says. "It would be nice to see a positive example of when a plane came to help them instead of to kill them. It would make them feel like they're not alone, that the world didn't forget them, and that there's still someone out there trying to help them."

Open Source

With all the parts prepped, the group clusters around the launcher on the lake bed. The Waliid sits on top of two metal rails that will guide it out and up while the bungee accelerates it forward.

The final verbal checks ring out while a small crowd forms to watch.

"Clear!" Jacobsen shouts.

The launcher releases and the drone springs forward. The bungee falls away as it coasts up and begins flying rectangles over Lake Lagunita.

Uplift flies its planes with software built by 3D Robotics. Jacobsen also runs a custom program that measures the plane's energy consumption at different flying speeds. Every so often, the Waliid increases its speed by 2 miles per hour, gradually moving from 28 to 50 miles per hour.

If Uplift begins sending drones into Syria, it will run another custom program. An app called Swarmify can take a single flight plan and turn it into as many semi-randomized flight paths as the team needs.

"Because every flight plan is slightly different, it ensures planes don't collide with each other," Jacobsen says. "It also gives you tactical survivability, because no two planes cross the same point on the ground. If somebody sees the first plane fly over, they won't catch the next one."

Much of the drone itself is made from off-the-shelf and open source components. While Uplift could someday manufacture its own drones, right now it works with inexpensive hobby kits. This choice has its roots in the organization's origins, when Jacobsen had to teach himself the basics of building and programming a drone and fund the project inexpensively.

But even as Uplift's volunteer ranks grew, it kept building its own drones. It turns out that there isn't much of an alternative.

"When we looked at different airframes, one thing became immediately clear: the market is really polarized as far as cost and capability is concerned," Fetroe says. "If you tried to put all the planes in a line, and had the tiniest, cheapest one on one end, and some huge commercial or military drone on the other end, you notice there's a really big gap in the middle, kind of where we are trying to operate."

Drones that can carry more than a few pounds of cargo for an hour and cost, say, $1,000, didn't exist. Fetroe said new options are emerging, but most have yet to officially hit the market. For now, Uplift will carry on with its own design.

The project's current hangar of vehicles cost between $500 and $1,000 to build. The Waliid is actually the $100 Talon kit made by X-UAV. Its autopilot system is built by 3D Robotics. Its motors, props, and servos were all picked for their modest price, and can be found on Hobby King.

In its belly sits the real value—the payload. A wooden box, laser cut by Fetroe, opens to release its cargo. It floats to Earth strapped to a parachute made from garbage bags, or whatever other cheap plastic is available (read more about this in Chapter 13).

وليد
١

Syria Is Not the Finish Line

Whether or not the Syria Airlift Project succeeds, Uplift sees a future for its drones. What will start with just a few flights this summer could scale to hundreds or thousands of planes that can feed entire neighborhoods. Even just a handful of planes can make rural medical deliveries and bring aid to disaster-stricken regions where the political situation is more welcoming.

"My long term goal is to help build a world where the use of starvation and medical deprivation are impossible—they just don't work anymore; you can always find a way to get humanitarian aid through. That's a lifelong ambition," Jacobsen said. "If we can get the first steps done, we can scale from there."

Quad Squad

5

These expert pilots and developers are working to make quadrotors cooler and more useful than ever.

Everyone loves a flying machine. Since launching just over a decade ago, DIY quadrotors and other autonomous aerial platforms have matured rapidly, thanks to an obsessive community and access to technology advancements like lithium-polymer batteries, brushless motors, and increasingly small, high-powered processors and sensors. With these components, drones are now incredibly strong, stable, and capable of doing most if not all of the piloting themselves.

So if these machines fly themselves, what do enthusiasts do to stay involved and excited? To help answer that question, we assembled a diverse gathering of top UAV flyers, including Hollywood filmmakers, smash-proof airframe builders, and aerial software and component creators, to discuss and demonstrate some of the newest tools and techniques involved in the pursuit of quadrotor aerial excellence. Their reports promise an exciting future in flight.

Going from 0-60 with APM

Figure 5-1 *Jason Short, Design Director, 3D Robotics, http://3drobotics.com*

APM:Copter was born on October 10, 2010. The date is indelibly etched in my memory, since it was the same day my son Lukas was born. We spent the week at the hospital while the Blue Angels flew overhead during Fleet Week in San

Francisco. I knew my days of flying UAVs at the airfield were likely over, so I set about designing one I could fly in my backyard while my son napped.

Adapting APM:Plane to fly multicopter drones was simple at first, but engineering full autonomy turned out to be a wicked problem. Multicopters stress the flight controller system. There are over 100,000 lines of code running on the Arduino-based processor, and almost nothing can go wrong that doesn't end in a bad outcome, often culminating in a crash. Producing a rock-solid flight control system and ironing out the details took a small army of volunteer developers and years of collaborative work, but the results have been astounding.

Today, the 3DR development team is focused on key features that will make it easier for new users to install and configure APM on any airframe.

Our latest software release of APM:Copter brings some new and very helpful capabilities. Setup wizards walk you through the configuration process, and a new auto-tune function learns how the drone flies, maximizing flight performance and removing the burden of manual tuning. A new, highly advanced inertial navigation controller fuses GPS and internal sensors to empower a pilot of any skill level to fly the drone right out of the box, without the challenges inherent to manual flight. Software-defined "geo-fences" prevent you from flying too far or too low. If the drone breaks the fence, APM automatically takes control and flies back home on its own.

A new flight mode called "drift" relies on the intelligence of the autopilot to simplify flight control to a single stick. The end result is a drone that flies and corners more like a race car than a typical multicopter. If you lose orientation, just let go of the stick and the brakes will be automatically applied, bringing your drone to a safe landing.

The most exciting improvement is our new, full-featured Android tablet interface, which enables you to plan and control a drone in the air. Community-developed apps like DroidPlanner and Andropilot allow you to command the drone with a simple Google Maps-like interface.

Advanced features, such as the "Follow Me" function, allow the tablet's position to be sent to the drone, creating your own personal flying camera, ready to capture your next hike up Kilimanjaro, surf in Maui or your son's first successful bike ride in the local park.

Building the World's Toughest Drone

Figure 5-2 *Marque Cornblatt, Cofounder, Game of Drones, http://gameofdrones.biz*

Deep in a huge Oakland, California warehouse filled with fire-breathing robots, monster machines, and other implements of destruction, a not-too-secret cabal of inventors, engineers, and artists meets late at night. This group gathers, first, to show off their latest custom-built drones, UAVs, and robots.

And, second, of course, to pit them against each other in one-on-one airborne "fights to the deck."

The crucible of destruction is known, somewhat informally, as "Flight Club." The first rule of Flight Club is that all commercially available drones and drone kits are far too fragile and expensive for heavy-duty use—especially if that use is dogfighting. But a number of innovative and perhaps even groundbreaking design concepts have evolved here, including many clever DIY methods for making drones cheaper, tougher, faster, and easier to repair.

Flight Club competition led me to team up with industrial designer (and long-time aerial-dogfighting nemesis) Eli Delia. Together we began researching highperformance materials and manufacturing methods from tough-duty industries including aerospace, military/law enforcement, and even medical manufacturing.

That research led us to thermoformed polymers, and we soon began designing and prototyping airframes using various sheet plastics including styrene, polycarbonate, PET, and Kydex 100, the supertough plastic "alloy" we ultimately settled on. Launching a Kickstarter let us test the market and get direct feedback from UAV pilots of all skill levels and needs, and this spark of user insight has already ignited several ideas for our next project.

My personal UAV—the one I fly every day—is one of our company's first prototypes. It's been crashed and/or dropped from hundreds of feet too many times to count. It has been flown through fires and landed in (and launched from) stagnant water. We've (deliberately) flown it through plate glass windows and shot it out of the sky with a 12-gauge shotgun. It keeps coming back for more.

Sure, it's scuffed, scratched, torn, and beat, but it still flies straight and true as the day we first launched it. The magic is in the airframe construction, and it's hard to imagine any other type that could withstand such abuse without becoming unflyable.

Besides the super-tough construction, we like to strip our airframes down to the bare necessities. For example, rather than using four ESCs on separate boards, we favor a four-in-one ESC board for motor speed control. This reduces the number of failure points significantly. The end result is a super-tough, super-simple airframe that can survive an entire day of flying, fighting, and crashing without a single repair.

Because most pilots go to great lengths to avoid collisions and crashes, most airframes—though they may be carefully designed to optimize other factors—are mechanically fragile. This has created a culture of expectation in which airframes that break when they crash are an accepted norm. Thus many amateur pilots are rightly afraid to take risks and really hone their flying skills for fear of damaging their frail, expensive gear.

At Game of Drones, our approach flies directly in the face of this culture. Our motto is "Fly 'em hard and put 'em away wet. They're only drones." It's my hope that this approach will not only make it easier for beginners to enter the hobby, but will also inspire more people to design, build, and fly drones for aerial combat games, business, research, and more.

Drones as Aerial Access Points

WiFi technology will make drones simpler to control and provide the opportunity, eventually, for Internet-controlled drones.

While the vision for drones is that they operate fully autonomously taking off, flying a mission and landing without human intervention, of course we will want to be able to find out where our drones are located, know whether they are functioning properly (or not), and be able to change the mission or take over manual control at any time. Achieving these ends will require maintaining wireless connectivity throughout the majority of each flight.

Figure 5-3 *Adam Conway, VP Product Management, Aerohive Networks, http://aerohive.com*

Wireless communications for hobbyist and pro-level UAVs today primarily consist of three connection uses.

Control

Steering a drone in manual mode, or switching into autonomous mode, is typically accomplished with a traditional R/C transmitter and receiver

Telemetry

As a drone is flying around it has the ability to send telemetry data back to a ground station. Telemetry data typically consists of onboard sensor inputs including GPS location and diagnostic data, but it can also be used to change settings on the drone mid-flight and provide new mission waypoints. Telemetry data is typically sent over a long-range serial link like IEEE 802.15.4.

Video

This is what gets most drone users excited—the idea of sending back real-time video so someone on the ground can experience what it is like to fly. For most hobbyists the only option for getting video from a drone is an analog wireless video transmitter/receiver. Analog video systems offer the advantages of being reasonably low-cost and having very low latency or lag.

With all three of these systems running at the same time there is a risk for interference (with potentially disastrous consequences), so most operators use different frequency bands for each system. Typically drone operators use 900 MHz for telemetry (at least in the US; 433 MHz is standard in Europe), 2.4 GHz for control, and 5 GHz for video. Since higher frequency means shorter range, video typically is the weakest link and will go out before an operator loses control or telemetry.

A better solution, however, may be to combine all three systems under a single wireless technology, one that has the range for flight but also the bandwidth to be able to deliver video, control, and telemetry with a single radio. For this, WiFi is the obvious choice: it's fast, inexpensive, and (if set up properly) has the necessary long range.

In the long view, WiFi and other TCP/IP-based networking technologies are going to be foundational for creating drones that are Internet-controlled.

Today there are already consumer drones, like the Parrot AR, that use WiFi for video and control signals. But among the more flexible open source autopilot software and hardware, support tends to fall off. However, a few eager engineers and hackers have already begun experimenting with adapting ArduPilot for WiFi telemetry and control, and I think it's only a matter of time until all drones move to WiFi.

Figure 5-4 *Jeffrey Blank and Andrew Petersen, Drone Dudes, http://dronedudes.com*

We are a unique collective of filmmakers, designers, and flying robots. Using a fleet of custom multirotor UAVs and custom camera gimbals, we offer our services as aerial cinematographers for feature films, commercials, music videos, and sporting events around the world. We feel fortunate to be supported by a network of amazing people and look forward to seeing where this exciting new technology will take our business and our art.

Our systems provide a cost-effective, safe, dynamic alternative to traditional aerial videography, making them an attractive substitute for producers considering conventional methods like manned helicopters and cranes.

Each UAV is designed with a different camera weight class in mind. Our heavy lifting octocopter was built to mount high-end cameras (like the RED Epic) that can produce the super high-resolution imagery the film industry now expects from professional camera operators. The RED camera, in fact, is the industry standard and flying it was our first big goal.

Now, with pro-quality HD cameras getting smaller and cheaper every day, we believe that the future of cinema drone technology is in a more compact system. Our new UAV design (the D2) comes equipped with everything a professional aerial video team would ever need for a shoot: onboard GPS, a custom three-axis brushless gimbal, full HD video downlink, wireless follow focus, and even dual parachutes for those "oh sh*t" situations. With great agility and response time, we expect the D2 to find a comfy spot at the top of the cinema-drone food chain.

We originally got into flying drones because they can capture shots that are not practical using any other camera platform. Now we've had a glimpse of what's possible, and are striving to constantly develop our technology. The complex, rapidly evolving intersection between technical development and artistic expression is what makes this business so much fun.

PART II
Tutorials

How can you do some of the amazing things people are doing with drones? The tutorial section contains four articles that can get you started.

The first two articles explain how drones are used to create beautiful examples of aerial photography and video—images that only a few years ago would have been impossible for the casual hobbyist to create. The third article shows how some enthusiasts are putting drones to work, creating aerial maps of historic buildings in Cuba. And finally, we learn some of the most important rules of flight, to keep drone use safe.

Getting Started with Aerial Video

6

How to make fantastic videos from your multicopter or drone.

—From Make:37 by Eric Cheng

The first aerial photograph was taken in 1858 by French photographer Gaspard-Félix Tournachon, from a hot air balloon. Since then, aerial perspectives in imaging have remained elusive to those without astronomical budgets. Historically, photographers have used just about everything to get cameras up in the air, including balloons, birds, kites, rockets, airplanes, and helicopters. In the last few years, unmanned aerial vehicles (UAV) have improved so much in performance and reliability that they have started to creep into the mainstream as the best way for (most) people to capture aerial images and video. These five tips will help you to get the best aerial videos you can.

Figure 6-1 *An Aerial view of Tonga, where most of this article took place*

Choose the Right UAV

The vast majority of people getting into aerial videography choose a quadcopter as their first UAV. Quadcopters' electronic flight controllers, sensors, and GPS automatically stabilize flight, and in some instances, allow autonomous "mission" flying via waypoint programming, allowing for steady video platforms that can maneuver themselves into precise locations. They're simpler to operate than tricopters, and more affordable than hexacopters.

The most popular quadcopter for aerial filming is the $679 DJI Phantom, because it's ready-to-fly (RTF) out of the box and is designed to hold a GoPro camera. The Phantom is a great platform, even for beginning hobbyists, because it's easily hackable. There is a vibrant third-party accessories market, mostly made up of enterprising individuals selling personally developed mods online.

Multirotors from 3D Robotics are also a great choice. They offer kits and RTF models (including a new Phantom competitor called the Iris), all running their open source, open hardware flight platform for the ultimate in hackability.

Figure 6-2 *The 3D Robotics Iris*

Adventurous makers will likely want to build their own multirotor aircraft, which have the advantages of being (potentially) more budget friendly (see Chapter 11) and allowing you to tailor components to your specific needs. A DIY quadcopter or hexacopter consists of an airframe, flight controller, electronic speed controller (ESC), motors, propellers, batteries, radio, and receiver. Entire kits are available for less than $200. Of course, to do videography, you'll also need a camera, which leads us to...

Choose the Right Camera

Although large cameras can easily be put into the air if you configure and make your own multirotor aircraft, my favorite cameras for aerial videography are GoPros, which provide the best image quality for their size and weight. The GoPro Hero3 Black Edition weighs only 73 grams and can record video at 2.7K (2,704×1,524 pixels) at 45Mbps (or 30fps). And it's got built-in WiFi for downloading your footage.

GoPros are also pretty much the standard in aerial videography, which means maximum compatibility with OEM and third-party accessories for aerial imaging, such as vibration isolators and gimbals (covered in the next tip).

Figure 6-3 *RotorPixel gimbals are matched to the DJI Phantom and also pretuned to match the GoPro Hero3 camera*

Finally, GoPros are easily protected while airborne using their branded underwater housing or third-party lens protectors.

Stabilize Your Camera

The smoothness of aerial video is directly correlated to its perceived quality. But multirotor motion isn't smooth. As a multirotor flies around, the flight controller automatically stabilizes the aircraft by sending power to its multiple motors. During flight maneuvers or in gusts of wind, a multirotor might pivot violently on multiple axes, which may keep the aircraft itself stabilized in space, but can wreak havoc on footage from onboard cameras. In the past, hobbyists used servomotors to correct for this sort of movement, but servos are slow and sloppy, unable to correct quickly enough.

Gimbals and Aircraft Motion

These days, stabilized aerial video is made possible by the incorporation of gimbals that use brushless motors. A gimbal is simply a support that allows the rotation of an object around an axis, and brushless motors are the same motors that revolutionized R/C model aircraft due to their great power-to-weight ratio (rewound for higher torque in gimbal use).

Figure 6-4 *Author's gimbal mount, showing the blue rubber vibration isolator*

A typical camera gimbal allows rotation around two axes: roll and pitch. A sensor on the camera mount tells the gimbal controller, "I want to be level," and the gimbal controller sends the appropriate signals to the brushless motors that control pitch and roll. In practice, brushless gimbals yield footage from quadcopters that looks like it was taken using a flying Steadicam (see *http://ech.cc/aerialvid* for some of my footage). Gimbals for GoPro cameras are available for as little as $150, and can simply be bolted to the bottom of any aerial platform.

Figure 6-5 *The aircraft is crooked but the camera is level, controlled by a brushless gimbal*

Prop vibration and "jello"

The second image-quality problem that needs to be solved is the removal of "rolling shutter" artifacts. CMOS image sensors, which are used in most digital cameras, scan the image in rows from top to bottom as they read data for each frame. If a camera is moved around during shutter sweeps, it results in horizontal spatial artifacts, more commonly known as "jello."

Jello is caused in UAV footage by high-frequency vibrations introduced by rotating motors and propellers. The best way to remove it is by balancing propellers, which can come from the factory with one side heavier than the other. Balancing is facilitated by inexpensive prop balancers, and is achieved by applying clear tape to the lighter side and/or sanding the heavier side. (Sand the flats, not the leading or trailing edges—YouTube has great tutorial videos.)

Balanced props, combined with the vibration isolators that are commonly used to mount gimbals, should yield beautiful, jello-free, stabilized video.

Figure 6-6 *Inexpensive prop balancers help you reduce propeller vibration*

Assemble an FPV System

It's difficult to get good video if you can't see what you're recording. With first-person view (FPV), an analog transmitter is used on the UAV to broadcast real-time video from the camera. The pilot uses a receiver and either a monitor or

LCD glasses to see what the UAV is seeing. Experienced pilots can fly 100% using FPV without needing a line-of-sight view of the aircraft.

An entry-level FPV system can be purchased for around $250. You can read my full deconstruction of the Ready Made RC 5.8 GHz starter kit at *http://ech.cc/quadfpv* —it taps into the GoPro to use it as the FPV camera as well. (For more details on using first-person view, see Chapter 8.)

Figure 6-7 *The author pilots his video Phantom over the waters of Tonga via an FPR (first-person view) system from Ready Made RC*

Practice, Practice, Practice

The most important thing you can do to improve your aerial video footage is to become a skilled pilot. There is no substitute for stick time, and spending all your time at a workbench instead of flying your UAV in an open field will never yield great footage.

Figure 6-8 *Practice your skills with toys like the Blade Nano QX and the Syma X1*

I recommend honing your flying skills using inexpensive off-the-shelf toys. The Syma X1 and Blade Nano QX or mQX are all great toy quadcopters that cost between $36 and $90. They fly using the same controls, and do not offer the luxury of GPS location hold. If you can master a small quadcopter, the skills you learn will translate directly to larger aircraft.

Figure 6-9 *More stick time = better video, so fly as much as you can*

Quadcopter Photogrammetry

7

How a trip to Cuba and my love of R/C aircraft aided in the restoration of historic buildings.

—From Make:37 by William Grassie

Nearing the end of my graduate program in media arts and computer science, I found myself stuck working on a thesis I no longer had much interest in. I had lost my motivation and feared I would end up in grad limbo with a project I couldn't bring myself to complete and expectations, including my own, unmet. About this time a friend of mine had signed up for a class that was going to Cuba. This was an opportunity I couldn't miss, so I signed up. This adventure led me to many others, including the genesis of what would become my new thesis.

I've long been an R/C flying enthusiast. In my boyhood, my dad and I built a small, gas-powered balsa wood plane. It was tethered to a string, and you could only fly in a circle. The poor plane didn't survive its maiden flight. That concluded my R/C experience for many years, as we couldn't afford to rebuild it. The price of all things R/C at that time made it cost-prohibitive for many.

A couple of years ago I discovered the hobby anew. I purchased a little R/C helicopter for my brother, and was surprised by the quality, flight time, and maneuverability. I started doing some research and found a whole new, more affordable world of R/C. This revolution was mainly due to the advent and proliferation of lithium polymer batteries and brushless motors, which replaced expensive, messy gas motors and made electric models a more realistic proposition. And so my obsession began.

Figure 7-1 *Grassie's first photogrammetry drone rig*

It started with small helicopters. Then larger helicopters built from parts. This led to airplanes, which was how I began doing FPV (first-person view) flying. Soon after came tricopters and quadcopters, which provided full three-dimensional freedom of movement and a very stable platform for cameras.

Then came Cuba, photogrammetry, and liberation from my uninspired thesis. Photogrammetry is a method for creating 3D models of objects by taking a series (usually hundreds) of photographs. The concept is as old as modern photography. What has changed is the use of digital photos and software. The software takes all of the photographs and compares them to find matching points. Then the software uses these points to calculate depth.

Through my graduate program in media arts at New Mexico Highlands University, I traveled to Cuba for photography, photogrammetry, and an exchange of ideas. One goal was to make contact with the Office of the Historian, which is responsible for restoring the buildings of Old Havana.

Highlands University had been working with the Georgia O'Keeffe Museum for about a year developing the use of photogrammetry as a tool for conservation and preservation. We hoped to share these simple and inexpensive techniques with members of the Office of the Historian. We contacted them, learned more about what they do, and demonstrated the methods we had developed for documenting historic objects and sites using photogrammetry. They were very excited and offered us the opportunity to create photogrammetric models of several buildings and structures.

This was my first real opportunity to use photogrammetry in the field, and I too was impressed with what was possible. However, while working on documenting several structures, it became apparent that we were limited by taking photos at ground level, which created gaps in the images. Once we had rendered a preliminary model of Hotel Santa Isabel, I found that anything above the field of view would inevitably show up in the data as black holes rather than a solid 3D model.

I started thinking of different ways to get a complete view of the building. One obvious method would be to rent a hydraulic lift, but that could be costly and impractical in tight spaces. Helicopters might work, but would also be cost-prohibitive. Then it hit me: I could use multirotor R/C aircraft to photograph the inaccessible areas. My passion for photography and the R/C world came together in a beautiful way.

Figure 7-2 *Photogramming the Hotel Santa Isobel*

Photogrammetry Tips

1. A digital camera with fixed focal length is best. 2. Make sure your photos overlap 60% to 80%. 3. Take the photos horizontal to your object and at a uniform distance. 4. Uniform lighting is important for creating good models. 5. Process the images in Agisoft PhotoScan. 6. Low-quality models can be generated on a laptop. High-quality models require multi-GPU systems with 128GB of RAM. 7. For small models you can get away with 30 to 60 photos. Larger models (like buildings) may require several thousand images. The more photos, the better detail you can achieve.

When I returned from Cuba I immediately got to work. I had only recently started experimenting with building quadcopters, and the one I owned had seen better days. But I went ahead and started modifying it to carry a camera for my proof-of-concept build. My initial test used a GoPro Hero HD set for time-lapse and my house as the subject. I shot 200 pictures, and the results, though not beautiful, were very encouraging. I set out to create a purpose-built quadcopter as a stabilized camera platform to create photogrammetric models of large-scale structures.

From the outset, I was determined it would be affordable and accessible, and I hoped my idea could inspire and educate others. In its simplest form, photogrammetry can be done with a compact digital camera and a laptop with surprisingly good results. But as the desired quality of the finished model goes up, the hardware requirements and processing time rise dramatically.

Figure 7-3 *The resulting 3D model of Hotel Santa Isobel*

Once I had built the new quadcopter, I began testing and collecting data. It worked flawlessly. I collected hundreds of photos to be processed and turned into a complete 3D model of the historic multistory building that was my subject. When the photos had been processed and a complete model had been created in software, I concluded that my methodology was sound and completely viable as a useful tool for photogrammetry of large-scale structures.

In association with the field-testing, I created a blog to help anyone who might be interested in getting started with their own quadcopter. The blog UAV 3-D (*http://uav3-d.info*) has articles on just about every concept of quadcopter flight so that this technology can be accessed by even the most uninitiated.

My Setup

MULTIROTOR: Custom built using parts largely from *http://rctimer.com* with an APM 2.5 controller board from *http://www.diydrones.com*. It has 30-amp SimonK ESCs (electronic speed controls) and 900kV motors with 10×4.7" carbon-fiber props.

CAMERA: Canon PowerShot running CHDK custom firmware, which lets the camera take RAW photos automatically.

FPV Fundamentals

8

Put a camera on your 'copter and yourself in the pilot seat.

—*From Make:37 by Steve Lodefink*

Watching your rotorcraft or fixed-wing R/C plane fly is always fun, but the experience really comes to life when you get to peer directly through the "eyes" of your aircraft, as if riding along inside it.

In R/C circles, this is called "first-person view," or more commonly, FPV. It refers to piloting a model aircraft from the perspective of the aircraft itself, via an onboard video camera, wirelessly linked to a ground station, streaming real-time video to be displayed on goggles worn by the pilot.

There are several ways to set up FPV on your rig; this guide should help you understand the general requirements and get you quickly up to speed.

Camera

The most popular cameras for FPV are small security-type "board cameras," which typically come as caseless circuit boards, with lens assemblies screwed right to the PCBs. It may be tempting to use a cheap, 480-line camera, but for a really satisfying experience, it's best to spend a bit more. $50 will get you a 600-line board camera from a trusted brand like Sony.

Not only will the higher resolution greatly improve visibility, but these slightly pricier cameras have dynamic exposure features that cheaper models often lack. Most importantly, "Wide Dynamic Range" (WDR) exposure compensation will allow you to see skies and shadowed ground features at the same time, without blown-out highlights or underresolved shadows. This is more than just an aesthetic concern; these features can make a big difference in your ability to navigate.

Besides "board" types, any number of small lightweight commercial video cameras could potentially be adapted for FPV drone use. As long as you can figure out power and signal connections, the only really critical requirement is low weight.

Mounting the Camera

As with shooting photos and video from your drone, it is especially important to keep airframe vibrations to a minimum when flying FPV. Vibrations cause blurry, nearly useless image transmissions. Balance all props, and if necessary, the bells of brushless motors. Mount cameras using foam, elastic bands, rubber standoffs, and/or other shock-absorbing means to soak up the shakes from the motors and props.

Ground Station

Your ground-based equipment is collectively referred to as a "ground station," and includes the video receiver, antenna, monitor or goggles, battery, and often a tripod, case, or backpack to house everything.

Ground station designs vary greatly. A good one will be easy to transport, quick to set up and take down, and difficult to trip over.

Frequency

A variety of low-cost miniature video transmitters and receivers are marketed with FPV in mind. Common frequencies include 5.8 GHz, 2.4 GHz, 1.2 GHz, and 900 MHz. There are several factors to consider when choosing a frequency.

1) Where do you live? Different countries regulate the radio spectrum in different ways. You may want to research your area's laws to avoid legal issues. 900 MHz has great obstacle penetration, for example, but may be reserved for phones, as in the US.

2) Where do you fly? Different frequencies have different characteristics. 5.8 GHz seems to have good range per watt, but is essentially line-of-sight and will not penetrate buildings or even trees. If you fly exclusively in open areas, 5.8 GHz might be a good choice.

3) What frequency do your controls use? Most R/C radios now operate at 2.4 GHz; to prevent interference, you may want to avoid that band for your video equipment.

Power

Most entry-level video transmitters radiate 100-500 mW. If you want to fly long-range flights, you may have to get a more expensive, higher power unit. Unlike airplanes, FPV multirotors tend to fly shorter-ranged missions, so you can still have lots of fun, even without a 10-mile range.

Antennae

The last thing you need when flying FPV is an unreliable video link. The little "whip" antenna that came with your video transmitter is useless. You will want to build or buy a better one. Three- or four-lobe omnidirectional "cloverleaf" or "egg-beater" antennas are a good choice and are easy to build and cheap to buy.

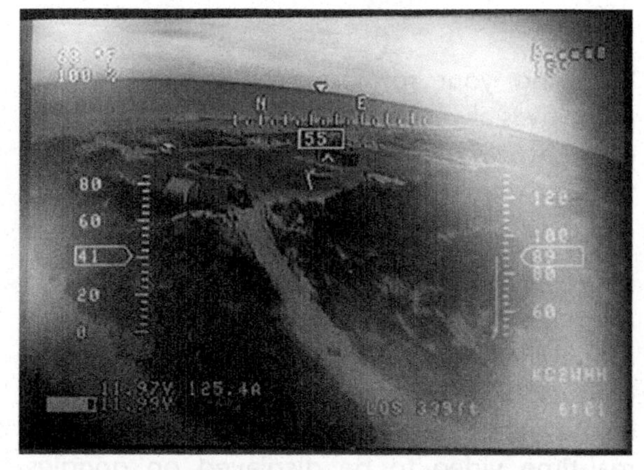

Figure 8-1 *A true heads-up display: video from drone, overlaid with flight information*

Many people also choose high-gain (but also highly directional) planar "patch" antennas for ground station use, and the best-equipped systems employ "diversity" setups that consist of two or more separate antennas, to get the best of both worlds. Special switching circuitry sends you the best available signal at any given time.

On-Screen Display

An on-screen display (OSD) is a little video processor installed in the signal path between the camera and the video transmitter. It takes information from its sensors and injects a graphical data display into the video stream. OSDs range in capability and cost, with the fancy ones featuring compass, GPS, barometers, telemetry, multiple battery voltage monitors, etc. While usually considered an "advanced" FPV system component, you can get a simple OSD for about $10 that does one very important thing: monitor and display the battery voltage. Knowing when you are about to run out of juice is pretty important for any FPV rotorcraft pilot.

Video Display

Some FPV pilots use an LCD monitor. I've found that piloting through video goggles makes for a much better, more immersive experience. There are video goggles made specifically for the FPV hobby, notably those by Fat Shark. Some models even have a video receiver built right into the goggles. I use a pair of MyVu Crystal goggles, which are general-use video glasses intended for watching video from an iPod, etc. I modified them by adding top and bottom shades of rigid foam, and a strap from a pair of sports goggles. As with cameras, you need to pay attention to resolution when goggleshopping. There's no sense using a 600-line camera with cheap 400-line goggles. As a rule of thumb, you'll want a pair with at least 640×480 resolution. If you use an OSD, chances are you won't be able to read the text on the display at a lower resolution.

Rules of Flight

8

Flying a UAV makes you a pilot, and like any pilot, you are responsible for the safe operation of your aircraft. The Drone Dudes, Jeffrey Blank and Andrew Petersen, share their rules of engagement.

- Know your equipment inside and out, and always double-check that everything is in perfect working order before each flight.
- Charge those LiPo batteries inside fireproof bags in a safe location with proper ventilation. Understand the hazards and science of LiPo battery charging, and keep an eye on the cell voltages yourself as you charge or discharge your batteries.
- Choose a safe fly zone away from buildings and highly populated areas. Think about what could happen if your aircraft fails mid-flight.
- Understand how changing weather conditions like temperature, altitude and wind will affect your overall flight performance.
- Check your onboard fail-safes and have a coordinated emergency plan with everyone in the flight area.
- Keep a safe distance from subjects and onlookers and always allow for unexpected drift from your plan.
- Keep a clear, safe zone for takeoff and landing.
- Make sure your payload is perfectly balanced on your airframe.
- Fly safe and stay alert. Listen to your gut and fly within your means. Do not let distractions divert your attention and don't hand the controls to anyone without proper training.
- Always fly line-of-sight so you can see what's going on. Do not solely rely on your GPS or flight controller to do the work for you. These tools can fail and you need to be prepared for that. If you are flying in an FPV mode (first-person view), use a spotter with binoculars to keep visual orientation of your aircraft for you.
- It's a good idea to always fly with a telemetry module that can relay live info about your aircraft. Watch your battery voltages for any irregular performance and keep your flight times modest, always flying on the safe side.
- Clear communication is essential. Make sure you have a reliable team supporting you and that everyone knows the predetermined flight.

PART III

Projects

Nothing helps humans to understand a topic better than doing hands-on projects about it. We close this book with a collection of project articles designed to help you get your hands around the reality of building and flying your own drones.

The first article shows how to create the Noodle Copter, a flyable quadcopter that is about as cheap and as sturdy as possible, since it is made out of foam! The second article shows how to build a completely autonomous drone, for approximately $1000. The third article covers the creation of the WAVECopter, a drone specialized for flying over water! The fourth project links back to Part One of this book—it shows how to create a payload mechanism to drop humanitarian supplies from a drone. And we end the book with an entirely new type of drone: a tricopter, which flies quite a bit differently from a quadcopter.

MENU
PAGE
EXIT
TARANIS
ENT

10 Noodle Copter

Here's a quadcopter that I designed and built in response to a CrashCast challenge to build a flyable quadcopter as cheap and as sturdy as possible.

—From Make:44 by Mark Harrison

Figure 10-1 *Structural supports for this drone are made from styrofoal pool noodles*

From a construction standpoint multicopters are interesting because (unlike helicopters) they have no moving parts and (unlike airplanes) do not depend on an aerodynamic body to fly. As a result, we see multicopters made from a wide variety of materials and construction techniques.

I had seen photos of a pool noodle unit previously and wanted to try one for myself for a few reasons:

- Pool noodles are cheap.
- It would be a good training unit when friends wanted to try flying. I can't imagine much you can to do break a pool noodle!
- It would be highly visible. My main quadcopter has really thin arms and is hard to see at a distance.
- It would be easy to light up for night and evening flying.
- I wanted to see how simply a working frame could be built. For example, the motors are simply taped to the frame.
- Let's face it, it's just funny to think of flying pool noodles!

If you build something like this, I encourage you to keep to the original spirit of the thing: improvise, have fun, and don't be afraid to try out new ideas!

Figure 10-2 *Pool noodle copter lit up with internal LEDs*

In keeping with the spirit of a pool noodle quadcopter, I wanted the construction to be as simple as possible (for example, the motors are taped to the arms). I bodged it together in an evening with materials that were at hand. While originally done as an experiment in minimal design and construction, I was happy enough with the results that I fly it regularly and use it as a trainer when somebody wants to try flying.

Step #1: Cut the Arms to Length

- Cut four pieces to 15.5 inches, and one piece to 5.5 inches. The four pieces will be the front and back arms (yellow in the picture) and the left and right side arms (red and green in the picture). The small piece will be the battery mount.

Step #2: Trim the Side Arms

- The side arms will be glued to the front and back arms, so we need the ends of these pieces to be curved. It will be easiest to do this with a jig. Get a piece of thin scrap plywood and cut a two-inch circle, using either a hole saw or coping saw. Cut across this hole so you have a semicircle at the edge of the wood.

- Take each of the side arms and the battery mount, and cut a semicircle out of the end of each piece. Be sure the ends of each piece are lined up. You should be able to dry fit the frame together with no gaps. I had to cut a couple of scrap pieces in order to get the feel for cutting the curve just right. You can compensate for small gaps while gluing, but if the gaps are too big try recutting the piece.

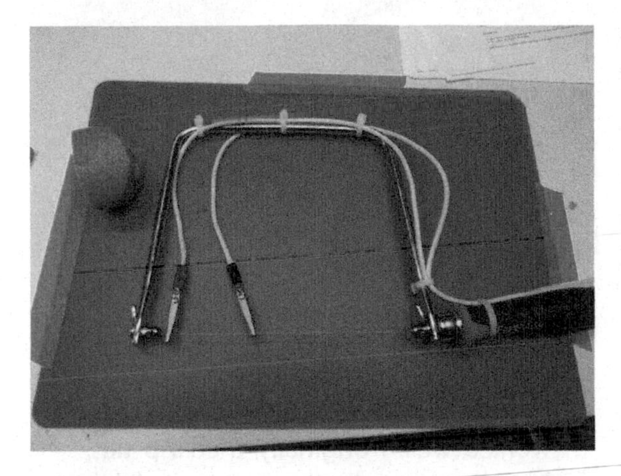

Step #3: Reinforce the Arms

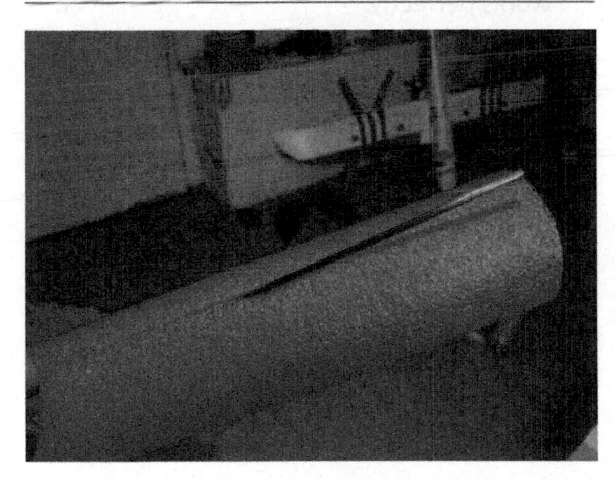

- Make a 1/4-inch slit along the top of each of the four arms. Be careful not to cut through to the center. Insert the 12-inch carbon-fiber rods into the slits on the left and right arms, and the 15-inch carbon-fiber rods into the slits on the front and back arms. I originally wanted to try 3/16-inch fiberglass kite rods, but I couldn't find a place to get them conveniently. Some pieces of split bamboo would probably work as well.

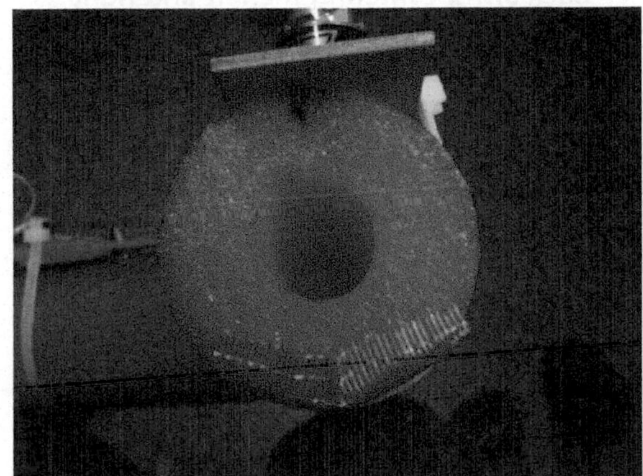

- Squeeze glue into the slits so that there are no dry areas in the slit or around the rod. This is important, as gaps in the slit will allow the arms to flex. Allow the glue to dry.
- I used Beacon Foam-Tac glue, and it worked really well. The only requirement for the glue is that it not melt the foam. Try gluing some scraps together if you're not sure.

Step #4: Attach the Arms

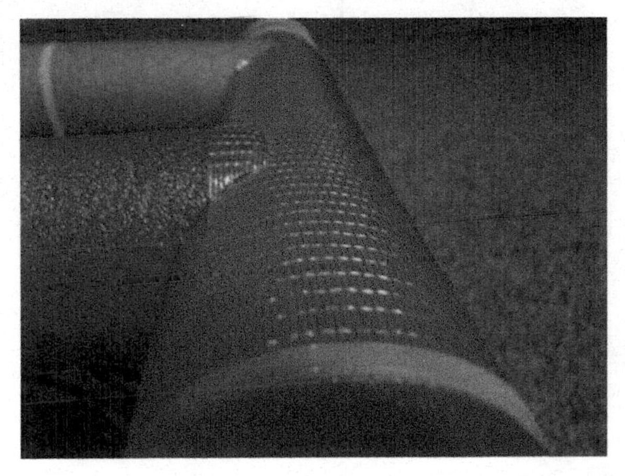

- Make marks on the front and back arms, five inches from each side. Glue the left and right arms to the front and back arms, centering the left and right arms on the marks you made on the front and back arms. Be sure you've got glue over the entire curved mating surfaces. Try to eliminate all gaps, but if you've got some small gaps it will be fine. Check your alignment, and secure with blue tape while the glue dries.
- Do the same with the battery mount. It fits centered between the left and right arms.

- The frame is complete after all the glue has dried. Check your joints, and make sure they're tight and strong. The frame should be relatively rigid (by pool noodle standards, at least).
- Run a length of strapping tape along the bottoms of all four arms: front and back, and left and right. Make sure it's tight and smooth. This, in conjunction with the rods in the top of the arms, will eliminate flex.

Step #5: Mount the Motors

- Cut four plywood scraps to a size that allows you to conveniently tape your motors to the arm. With my roll of tape, a dimension of 1 inch by 3.5 inches worked well. I originally tried zip ties, but they cut into the arm and didn't keep the motors level. Strapping tape works perfectly.
- Tape the four motors to each end of the front and back arms. Use strapping tape on each side of the motor. Run the tape all around the arm so it overlaps with itself. Make sure it's smooth and that the motors are flat on top of the arms. You can adjust the motor position by wiggling them a bit.

Step #6: Configure the Center Mount

Cut out some of the bottom battery mount. There should be just enough room for a 3S 2200 mAh battery to fit. I bodged the flight control board mount by hot gluing the electronics to a piece of scrap coroplast plastic (just about anything flat will work), and then attaching the coroplast to the top of the battery mount with two velcro straps, one on each side. A third velcro strap holds the battery in place.

Step #7: Flight Electronics

- Configuring the flight electronics is the same as for any other quadcopter. You can follow Chad Kapper's excellent instructions here (*http://makezine.com/projects/make-37/the-handycopter-uav-2*). The Ardupilot Mega is my preferred flight controller board, but I've tested the KK2 and Flip controllers and they also worked well. Note that everything is taped, hot-glued, or zip-tied to the top of the unit. I did this because I wanted to finish it quickly, and I honestly thought I would recycle the parts after my friends and flying buddies had gotten a good laugh.

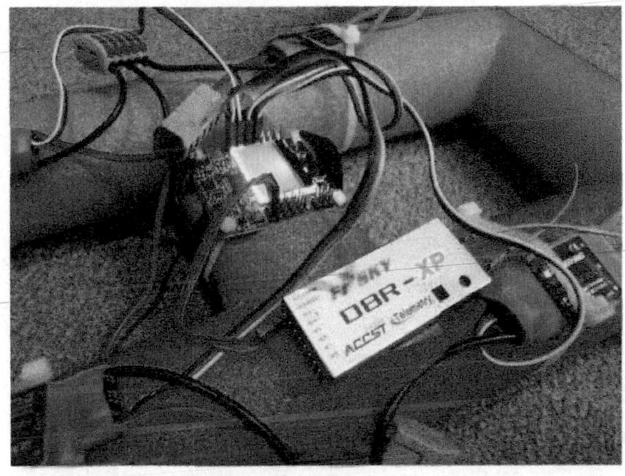

If I make another frame, I'll try running some of the wires, etc., either through the arm or embedding them like the rods. Interestingly, while it's incredibly ugly viewed from the top, it looks quite nice from the air since you're mainly viewing the bottom. The farther away it is, the nicer it looks.

- I didn't have a power distribution board handy, so I used Wago connectors to do this. Here are some notes on doing this: *http://eastbay-rc.blogspot.com/2011/03/update-wago-connector-for-power.html*.

- I make a big mistake though. I cut a slot in the arms for the LED batteries. They were a perfect fit, but the cut in the arm allows the arms to twist, reducing the rigidity of the frame. I taped a couple of extra carbon-fiber rods to the plywood motor mounts in an X shape to counteract this, but if you don't cut into the arms that won't be necessary.

Step #8: Adding Lights

- Lights were simple to add. LED strip lights fit perfectly into the hollow pool noodles. I used a separate battery for the lights so I wouldn't have to run an extra wire to the main battery.

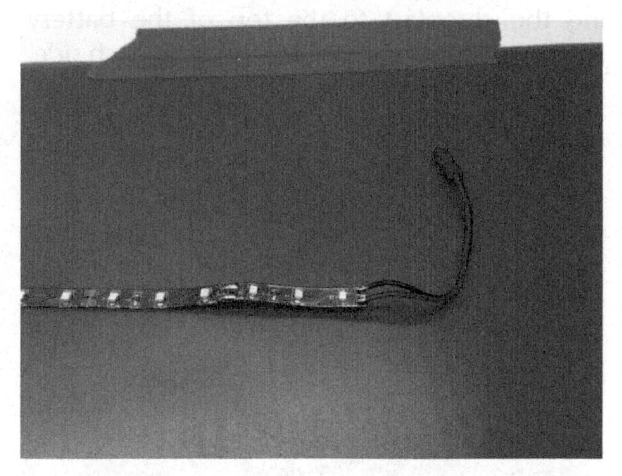

- At night the yellow arms were quite a lot brighter than the red and green arms. I arranged it so that the yellow arms were powered by a 2S battery and the red and green arms were powered by a 3S battery.

Step #9: Test Flights and More Information

- The Noodle Copter flies quite well. The fat arms catch the wind more than other copters, but it's stable even in pretty high winds. We maidened it with gust of wind up to about 15 MPH: *http://bit.ly/1WO7qSI*.

- Here's a few more articles on the Noodle Copter. You can get more information on the Noodle Copter here: *http://bit.ly/1q5S3e0*, *http://bit.ly/1T9mhrQ*, and *http://bit.ly/1pLSV7T*. You can follow the progress of the Noodle Copter and other East Bay RC projects here: *http://eastbay-rc.blogspot.com* and *http://bit.ly/1pLT6zS*. There are a lot of tutorials on making and flying RC craft of various sorts, and on using the various electronic bits that make model aircraft fly these days.

11 The Handycopter UAV

There are essentially two configurations for a quadcopter: the "+" frame and the "X" frame. Here we've chosen to build an X frame so your onboard camera can have a clear forward view.

—*From Make:37 by Chad Kapper*

We'll take you all the way from building the airframe to adding autonomous flight capability with ArduPilot. Once you've got it working, you could program this drone, for instance, to automatically visit a series of landmarks or other waypoints and take pictures of them.

Time Required
A weekend

Cost

Airframe
$30–$60

Avionics
$500–$800

Materials

For the airframe:

- Conduit clamps, 1 1/2" (4)
- Square dowels, wood, 1/2"×36" (2)
- Machine screws: flat-head M3×6mm (8); M3×20mm (12); M3×25mm (4)
- Hex nuts, M3 (8)
- Flat washers, M3 (4)
- Thread-locking compound
- Liquid electrical tape
- Polycarbonate sheet, 0.093"×8"×10"
- Zip ties, 4" (100-pack)
- Flexible PVC coupler, 1¼" to 1¼"
- Aluminum bar, 1/8"×3/4"×36"
- Hook/loop strap, 1/2"×8" (2)
- Hook/loop tape, 3/4"×18"
- Weatherstrip tape, foam, 3⁄8"×12"
- Double-sided tape, 1"×5′
- Wire, stranded insulated, 12 AWG, 12" red and 12" black

For the avionics:

- Copper pipe reducer, 1" to 1/2"
- Gimbal motors (2), iPower 2208-80
- Gimbal controller, iFlight V3.0
- Flight controller, 3D Robotics ArduPilot Mega 2.6
- GPS module, 3D Robotics LEA-6H
- R/C transmitter, 5+ channels
- R/C receiver, 5+ channels
- Motors, 850kV (4) AC2830

- Propellers, Turnigy 9047R SF (2)
- Propellers, Turnigy 9047L SF (2)
- Electronic speed controllers (4)
- M/M servo leads, 10cm (5)
- Camera, GoPro Hero3 White Edition
- LiPo battery, 2,200mAh, 3S 20C
- Battery monitor, APM Power Module with XT60 connectors

Tools

- Computer with printer
- Straightedge
- Plastic scoring knife
- Drill and bits: 1⁄8", 3⁄16", 1/4", 5⁄16", 3⁄8"
- Wood saw
- Phillips screwdrivers: #1 and #2
- Pliers
- Wire cutters / strippers
- Hacksaw
- Soldering iron and solder
- Scissors
- Pencil
- File
- Hobby knife

Figure 11-1 is a look at the overall wiring diagram for the Handycopter UAV. We'll then take you step-by-step through the process of creating it.

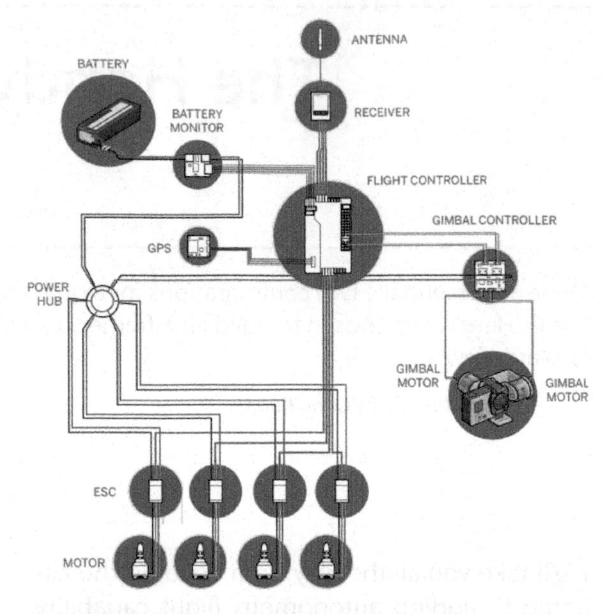

Figure 11-1 *Wiring diagram for the Handycopter UAV*

Step #1: Fabricate the Body

- The copter's central hub consists of two polycarbonate plates. Download the cutting and drilling templates from *http://makezine.com/the-handycopter-uav*, print them full-size, and affix them temporarily to your polycarbonate sheet.

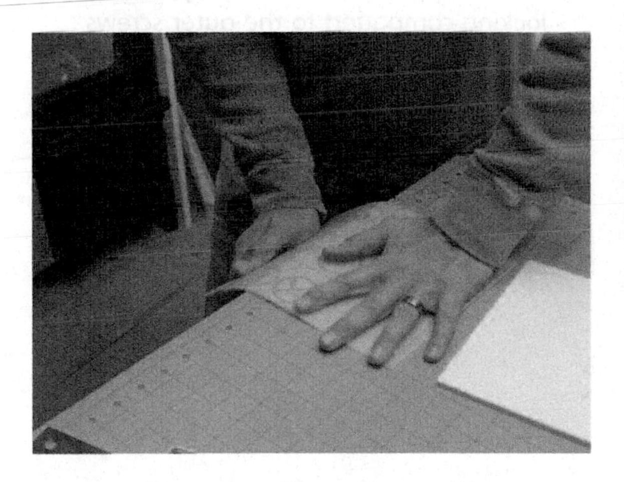

- Use a plastic cutter to score and snap each plate to shape, then drill out the holes with a 1/8" bit.

Step #2: Cut and Drill the Booms

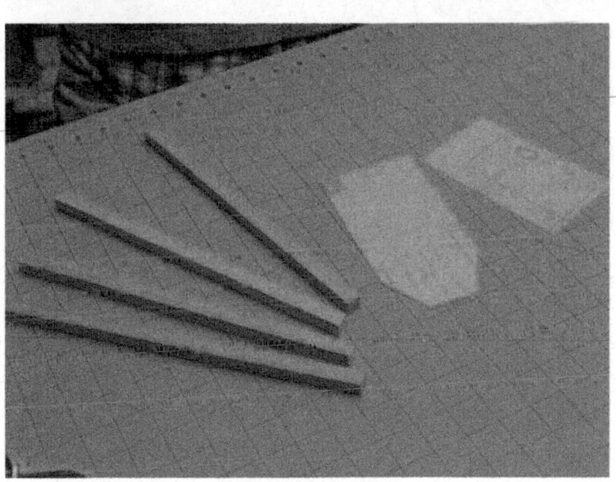

- Saw four square dowel booms to 10"–11" each. Shorter booms will make your quad more agile, and longer booms will make it more stable.

- Drill two 3mm holes, one 6mm and one 26mm (on-centers) from the end of each boom.

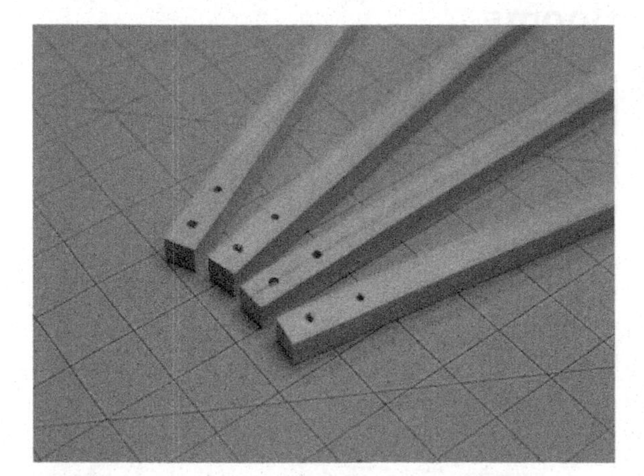

Step #3: Assemble the Frame

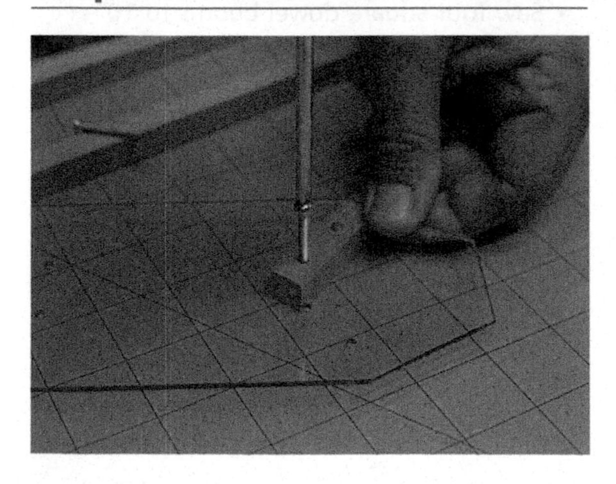

- Secure the booms between the hub plates using four M3×25mm screws through the inner holes and four M3×20mm screws through the outer holes.

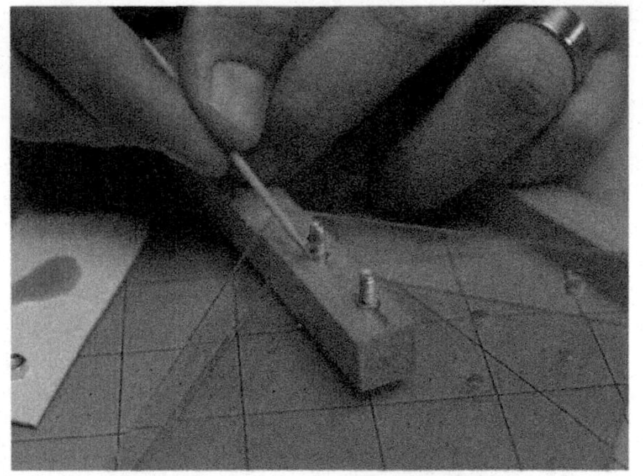

- Once the booms are in place and you're happy with the fit, apply thread-locking compound to the outer screws only, add nuts, and tighten them down.

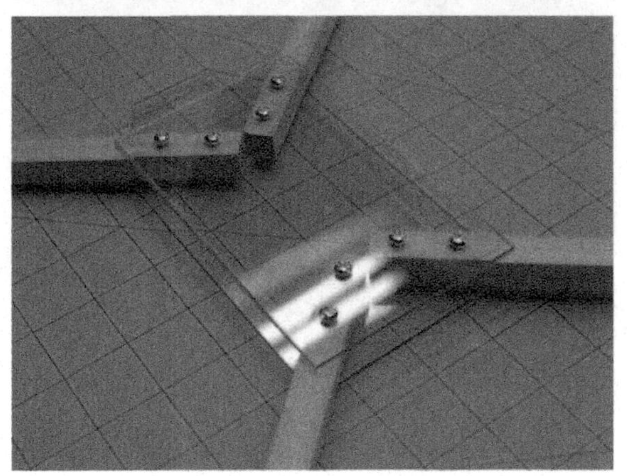

Thread the inner nuts on just loosely, for now.

Step #4: Wire the Power Hub

- Six components will connect to the power hub—the four electronic speed controllers (ESCs), the power module, and the gimbal controller board.

- First, cut off the male XT60 connector from the APM power module cable. Then strip about 1/4″ of the insulation from each wire, red and black, on all six components, and tin the stripped ends. Saw a 3/8″ ring from each end of the copper reducer, and file off any rough edges.

- Solder each of the six red positive leads to the smaller ring, and the corresponding six black negative leads to the larger rIng. Wrap the smaller ring in 3/8″ foam weatherstripping tape and slip the outer ring over it.

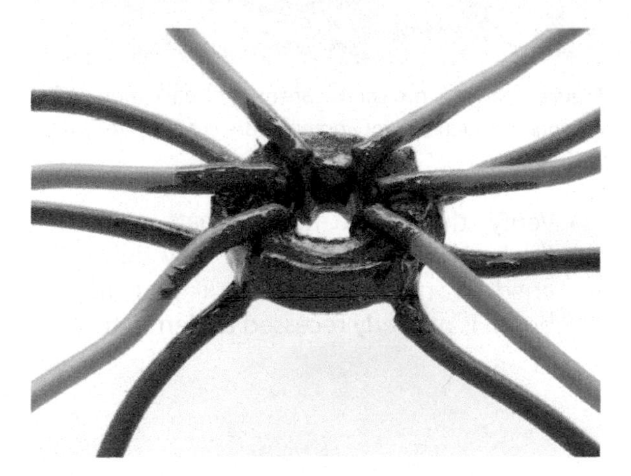

- Finally, paint the entire hub with liquid electrical tape for insulation.

- Wiring the motors and electronic speed controllers together is tedious. Store-bought distribution boards are convenient, but cost space and weight. I prefer this homemade distribution hub made of rings of nested copper pipe to keep things lean and tidy.

Step #5: Drill Motor Shaft Clearances

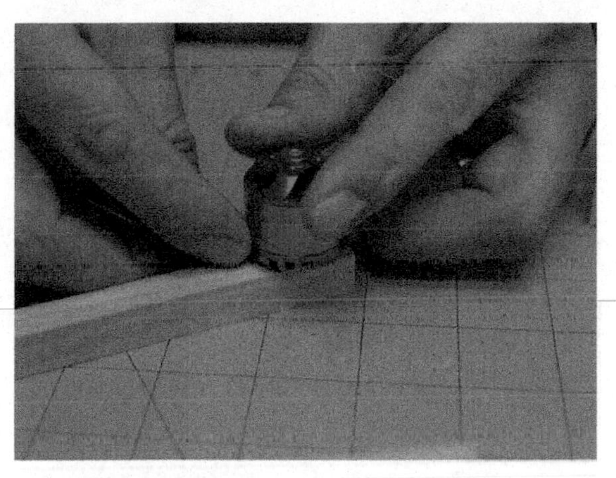

- Here we'll show you how to make your own landing struts from ordinary conduit clamps. You can also use inexpensive prefab combination landing gear / motor mounts that simplify the process quite a bit, and look better to boot. Please check out our product line at *http://www.flitetest.com* if you're interested in the prefab option.

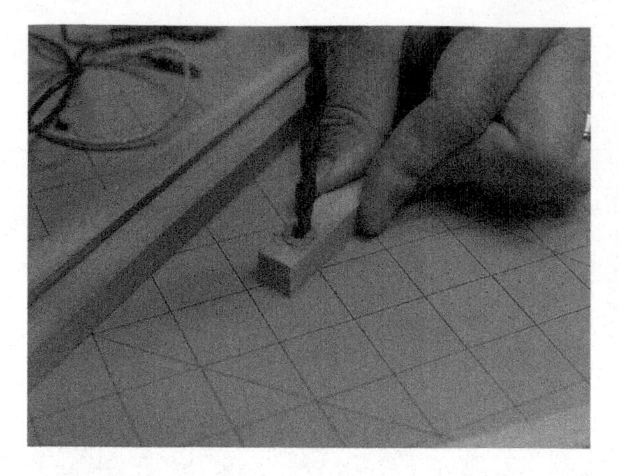

- If you go the homemade route, you'll be mounting the motors directly to the booms. Mark and drill a shallow blind recess in each, so the shaft can spin freely. A 5⁄16" bit works well for this.

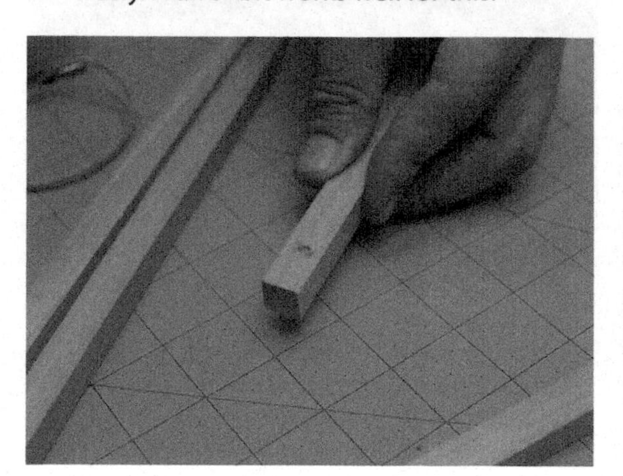

Step #6: Mount the Motors

- Cut down the bracket that came bundled with each motor and use two M3×20mm screws to clamp a motor to the end of each boom.

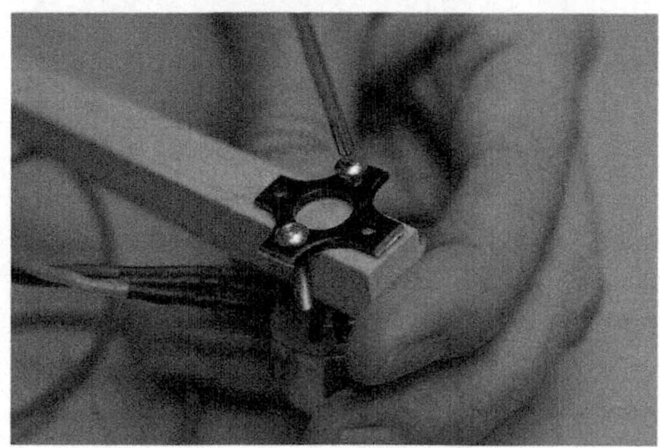

Figure 11-2 *Only two screws are used on each motor for mounting, and the factory brackets are cut down to save weight*

- Verify that each motor shaft spins freely when the screws are fully tightened. If not, double-check that its boom is properly recessed underneath.

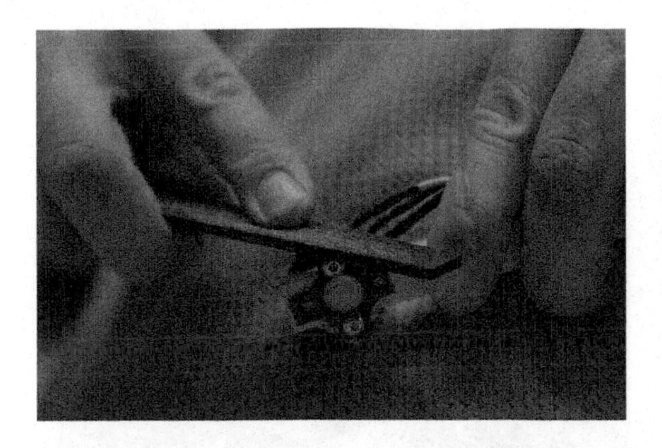

• Smooth any rough edges on the bracket with a file.

Step #7: Add the Landing Gear

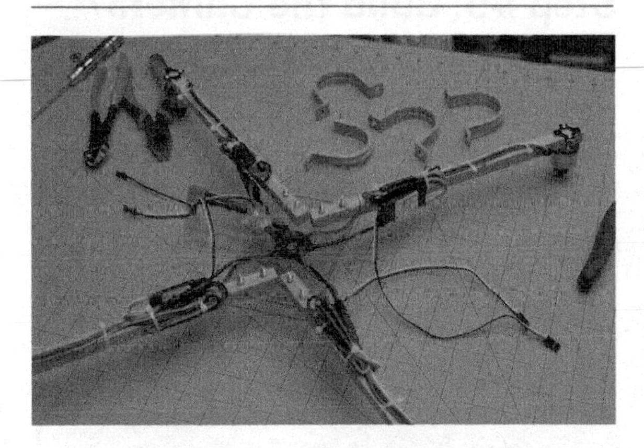

• Slip the power hub between the top and bottom body plates and route the ESC power leads out along the four booms. If you bought motors and ESCs from the same manufacturer, there's a good chance they came with preinstalled "bullet" connectors. In this case, simply plug the motor leads into the ESC leads and coil any slack under the boom. Or you can solder the motor wires directly to the ESC boards for a cleaner build. Secure the motor leads, the ESC power leads, and any leftover slack tightly against the booms with zip ties.

• Use wire cutters to snip off one side of each of four conduit clamps, leaving a J-shaped foot behind. Smooth the cut end with a file, then file or grind two small notches beside the remaining mounting hole as shown. Attach a foot to the end of each boom, just inside the motor mount, using a zip tie run through these notches.

Step #8: Install the Shock Mounts

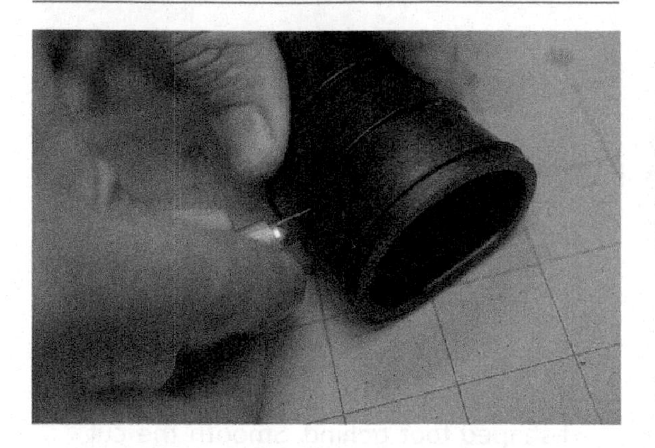

- Remove the hose clamps from the flexible PVC coupler and save them for another project. Cut two 3/4″ rings from the coupler's rubber body with a sharp hobby knife. Align each ring across two of the frame's protruding inner screws and press down hard with your thumbs to mark two drilling spots.

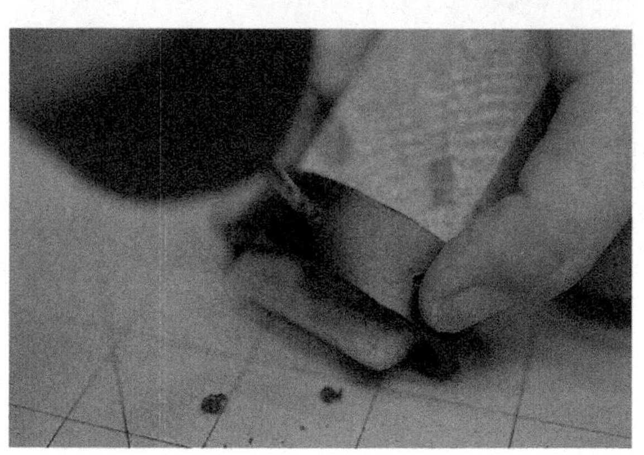

- Drill 1/8"-diameter holes on the dents, through one side of the ring only. Install the rings over the frame screws with M3 flat washers and nuts. Secure with thread-locking compound when you're happy with the fit.
- The gimbal and battery shelf are attached via two shock mounts cut from thick flexible rubber tubing, which helps isolate the camera from propeller vibrations and adds a bit of space, above, to mount the gimbal controller board.

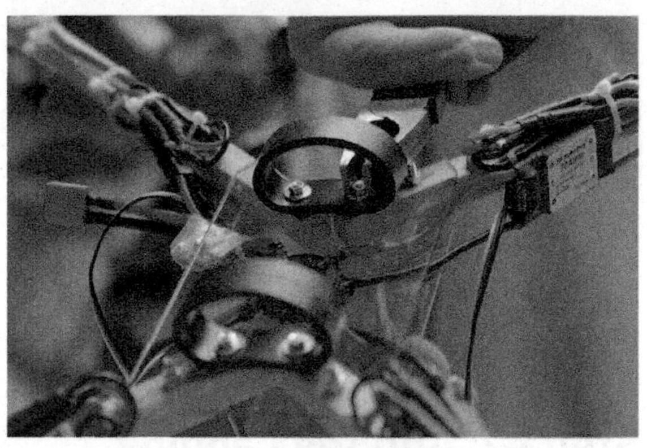

Step #9: Build the Camera/ Battery Mount

The gimbal and battery shelf are assembled from three simple L-shaped brackets. We refer to these as the shelf, roll, and pitch brackets.

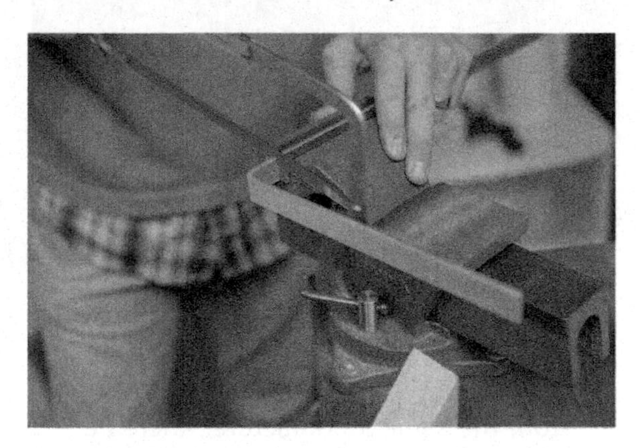

- Saw a 36″ length of 1/8"×3/4″ aluminum bar stock into two 18″ sections, then saw one of those into two 9″ sections, giving three pieces total. Make a right-angle bend in each section as indicated on the templates, working over a piece of wood or other scrap with a beveled edge to increase the bend radius to about 3/8″. (Too sharp a bend can overstress and weaken the aluminum.) After you've made the bends, cut each bracket to final size per the templates.

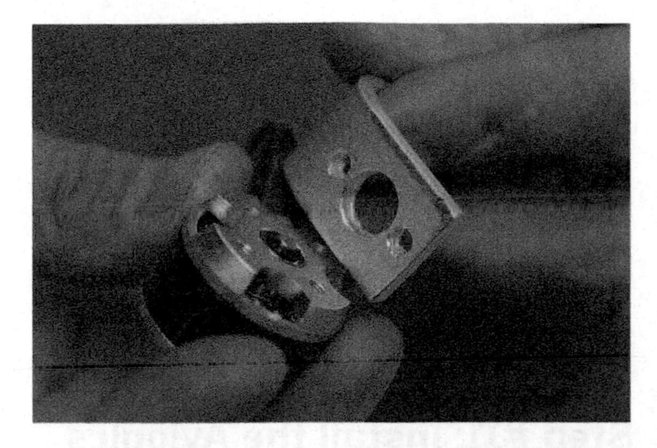

- Accurately locate, mark, and drill a centered row of three 1/8"-diameter holes on the short leg of the shelf and pitch brackets, and on both legs of the roll bracket. In each case, the outermost hole should be 3mm from the bracket end on-center, and the holes themselves 9.5mm apart on-centers. Finally, step-drill the center hole in each row up to 3/8" to provide clearance for the motor shaft.

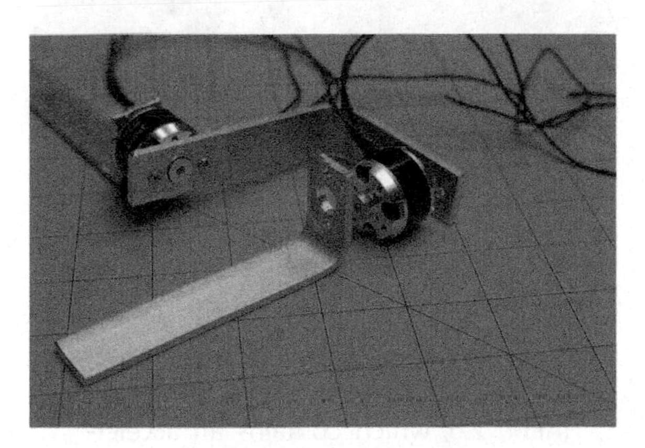

- Use two M3×6mm screws to attach the bottom of a gimbal motor to the shelf bracket, and then two more to attach the top of the motor to the longer arm of the roll bracket.
- Attach the bottom of the second motor to the free arm of the roll bracket, and its top to the pitch bracket, in just the same way.

Step #10: Mount the Camera and Battery

I designed this quad to balance properly with a 3S 2,200mAh LiPo battery and a GoPro Hero3 White. If you use other equipment be sure you keep the CG (center of gravity) in the middle of your airframe. Here's how to get it balanced.

- With the frame upside-down, balance the camera, brackets, and battery across the two shock mounts on the underside of the frame. Adjust the position of the whole assembly forward and backward along the frame until the entire quad balances evenly between your fingertips, centered on either side of the body.

- For the gimbal motors to operate smoothly, the camera must be balanced along both axes. Weaken the adhesive on a piece of double-sided tape by sticking it to your shirt and peeling it off. Remove the backing and apply the exposed side to the pitch bracket, then use the weakened side to hold your GoPro in place while you adjust it to find the balance point. Once you've got it, use an elastic band or a velcro strap, in addition to the tape, to hold the camera securely in place.

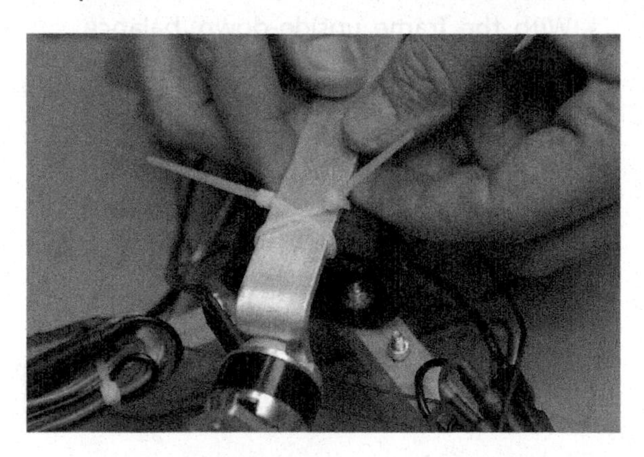

- Once you've got the CG right, fix the shelf bracket to the shock mounts with two sets of crossed zip ties. Apply hook-and-loop tape on top of the shelf bracket and on the underside of the battery, and fix the battery in place. Add a hook-and-loop strap around both bracket and battery as an added precaution.

TIP

Though the GoPro is a tough camera, you may want to build a "dummy" version having the same weight, and approximately the same size, to mount during your maiden and subsequent shakedown flights.

Step #11: Install the Avionics

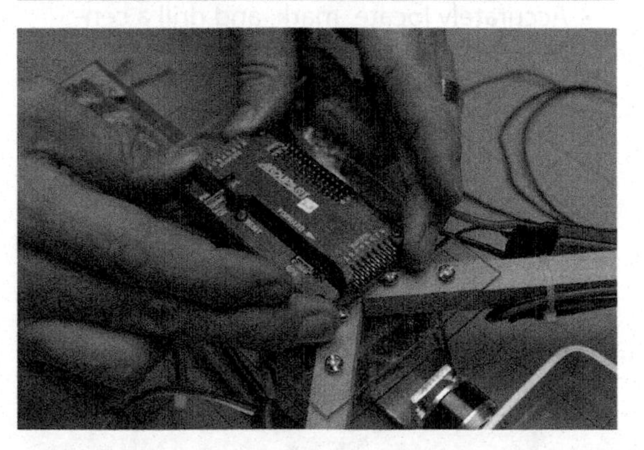

- Arrange your flight controller, receiver and other modules before attaching them to the airframe. Once you're happy with the layout, use double-sided tape to secure everything to the frame. Download the wiring diagram for a detailed list of all connections.
- Attach the flight controller. In this build we use 3D Robotics' ArduPilot Mega (APM) 2.6, which contains an accelerometer and must be oriented correctly with respect to the frame. Align the arrow on the APM case toward the front of the quad and fix it in place with double-sided tape.

- Add the GPS/compass module, which fits neatly on the rear extension of the bottom frame plate, and also must be aligned with the arrow forward. Tape the module in place and connect the cable to the APM's "GPS" port.
- Starting from the starboard-front position and proceeding clockwise (viewed from above), connect the ESC signal cables to APM outputs 1, 4, 2, and 3.
- Mount the receiver alongside the APM with double-sided tape, and connect channels 1–5 to the corresponding inputs on the APM.

Step #12: Install the Gimbal Controller

The gimbal controller consists of two boards: the larger controller board and the smaller IMU sensor unit. The controller board goes above the shelf bracket, in the space provided by the shock mounts.

Cover the top surface of the bracket with foam weatherstripping to keep the solder points from shorting against the bare aluminum, then fix the controller board to it with zip ties. The IMU detects the orientation of the camera and needs to be mounted in the same plane; fix it to the underside of the pitch bracket with double-sided tape, and run the connector cable back to the control board. Connect the three wires from each gimbal motor to the ports on the controller. Secure all wires with zip ties, leaving plenty of slack for the gimbal to rotate freely.

Step #13: Configure the Software

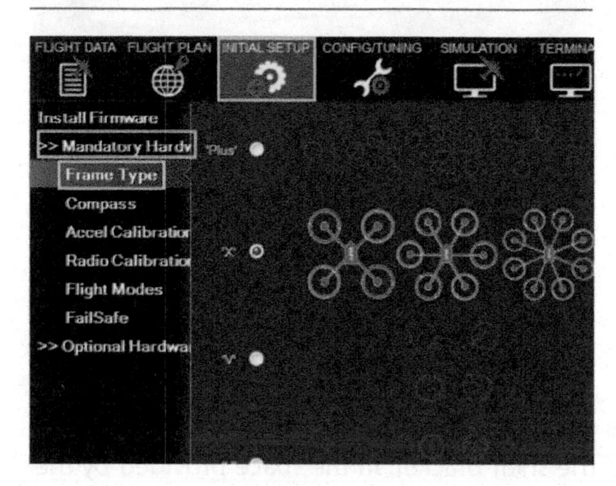

- The flight controller, ESCs, and gimbal controller all need to be calibrated and configured before flight. Refer to the bundled or online instructions that came with your equipment. Specific tutorials are available through *http://makezine.com/the-handycopter-uav*.

Step #14: Add the Props

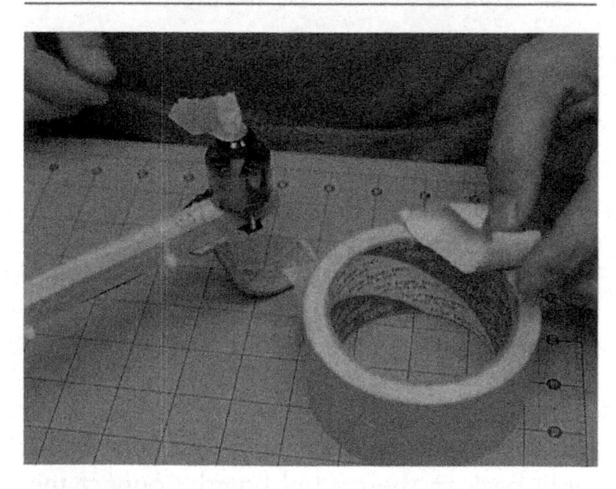

- Before you install the propellers, put bits of masking tape on the motor shafts to make it easy to see which way they are spinning. From above, motors 3 and 4 should spin clockwise, and motors 1 and 2 counterclockwise (see Chapter 3). If a motor is reversed, simply swap any two of the three leads connecting it to the ESC.
- The most important factor for steady flight is balanced props! There are lots of tricks for doing this, but the simplest involves sanding the heavier side of each blade until the prop balances level on a horizontal shaft. (Sand only the flat, not the leading or trailing edges.)

- Once the props are balanced, install them on the shafts and tighten the nuts. You'll use two conventional airplane "tractor" props and two reverse-pitched "pusher" props. Motors 1 and 2 take tractor props, and motors 3 and 4 take pusher props. (If you're not using the APM flight controller, your prop configuration may be different.) Once you've got it right, mark the number and direction of rotation for each motor on its boom for easy reference.
- Make sure the props are balanced, the parts are securely fastened, and none of the props, gyros, or controls are reversed.

WARNING

If you need to reverse a motor, be sure to swap the motor control leads only, not the ESC power hub leads. Don't ever reverse the power connections on an ESC!

The Maiden Flight

Verify that all your radio trim settings are at zero (if you have to trim, do it through the APM, not the radio). Wait for wind-free conditions to actually make the first flight.

Conclusion

Don't expect your quad to fly perfectly the first time. You'll likely need to make some tweaks and adjustments before it flies well. If you've never flown a quad before, remember to work the controls gently, as most beginners tend to oversteer. Your first goal should be to hover about 24" off the ground for 1-2 seconds and then immediately land. Once you can do that consistently, try to take off, rise above the "ground effect" zone (3'-4'), and then land gently. Work your way up gradually to longer and higher flights.

It is likely that you will crash at some point, especially if this is your first multirotor. Keep a positive attitude, pay attention, and try to learn something every time. Crashing, learning, repairing, and improving your skills and your machine is part of the fun and challenge of the hobby.

WAVEcopter: A Waterproof Quadcopter

12

WAVEcopter is a fully waterproof multicopter frame that I have constructed mostly from readily available and cheap electrical parts. My reasons for building it were to gain a new perspective on surfing photography, to do aerial surveying of event sites, and to satisfy my general fascination with robotics and aviation.

—From Make:44 by Alec Short

I rebuilt the copter after it suffered a serious collision into a cliff on the south coast of Cornwall, England. (I believe it was a pilot error after relying too much on a GPS fix and moving the copter from its initial takeoff point.) Although this was a serious collision—full throttle into a granite cliff face at about 60 feet—all of the electronics and camera equipment were unscathed. This was testament to the ruggedness of the airframe; damage was limited to broken props, a cracked hub, and severed motor wires, all easily replaced for minimal cost.

The copter is now on build 2 as I'm planning on replacing the Naza M flight controller with an ArduCopter control board to enable mission planning with waypoints.

There are lots of different setups for copter electronics, so I've skimmed over some of that detail as I think there are better setups than mine (like the ArduCopter). The most important part of this project is being able to waterproof the frame while housing an optimum balance of battery power and weight.

The drone has made successful flights of over 10 minutes with no apparent overheating of speed controllers or motors, which was a big initial concern in a airtight/waterproof frame. I'm new to multirotors and I've read a lot about this, so it's either a myth or I've been lucky so far. You can add heatsinks for the electronics on the underneath of the main housing if you're overly concerned.

I hope you enjoy the project!

Step #1: Prepare the Hub

- Place the 4-way PVC intersection face-down inside the weatherproof electrical socket box. You'll have to trim the ends with a hacksaw to ensure a snug fit.

- Remove the hinged lid from the main box to make things easily accessible. Pop out the center lugs and make sure the 4-way intersection aligns with the holes. It should do perfectly.

- You can see I cut out holes in the base of the box to attach heatsinks, but I ended up not using them in the next build as they didn't seem to get hot.

Step #2: Prepare the Rotor Arms

- Cut the 1000mm tube into four equal lengths of 250mm. I used a hacksaw.
- Take all four conduit reducers and grind away the little lip you can feel on the inside, so it's flush with the internal diameter. I just held them in my hand and used a Dremel with a drum sander attachment. Make sure you can now slide the carbon tube through the reducer.

You might find you have sanded too much—if so, use another reducer. The tube needs to slide through with a good push and no less.

- Now lightly tap the reducer into the 25mm thread adapter until it stops. Unscrew the lock ring from the adapter and poke the threads through the main box, then screw the lock ring back on to form a tight connection. Do this for all four connections.

- Now take the carbon tubes and push them/tap them with a rubber mallet through the thread adapters and into the 4-way hub as far as you can. Make sure each arm protrudes from the box an equal distance as the opposite arm (within 1mm–2mm).

Step #3: Make the Motor Mounts

- Cut both ends off the 3-way inspection tees, as far back as you can. Make sure the cuts are square. Depending on your motors you might have to glue a tap (faucet) washer on top of the inspection cover to give the motor screws enough clearance. (I trashed a motor here by screwing into the windings, so be careful.) The washer also acts as a vibration cushion for the motor.
- It's very important to upgrade the screws that come with the inspection cover as the motors will pull them free. I used some M4 machine bolts and just drove (forced!) them through the threads. It's probably smarter to use smaller bolts (M3) with nuts.

- Attach the rubber grommets to blank off the cuts you made earlier. This is a bit fiddly but they do fit; use a small flat-blade screwdriver to pry them in. They make an excellent seal once in.
- Attach the motor mounts to your rotor arms by tapping slightly until they are level and tight. You may have some slight movement or twist on the mounts that may seem alarming but we can fix this later with some PVC pipe welding cement.

- Make a hole in the inspection cover just big enough to thread the motor wires through. You can seal this later with Sugru.
- Now that everything is fitting together and you're happy—take it all apart, as you're about to install the electronics.

Step #4: Install the Power Electronics

- As there are many different ESC/motor/flight controller options out there, I will assume the reader has a basic understanding of how these are wired up. It's pretty simple.
- For the wiring harness of the 30A Naza ESC, measure the distance from the motor mounts to the center of the box and add 300mm. Cut your silicone wire and begin soldering the bullet connectors that you'll use to connect the power distribution board. (You don't have to use one of these boards, but I found it saved a lot of time.)

- It's a good idea to number or color-code the ESCs at this stage because you're going to be stuffing a lot of wires and you can easily get confused when you're calibrating the flight controller. I used little coloured stickers for a reference.
- Reattach your rotor arms and motor mounts and drill a 25mm hole on top of the 4-way hub. Thread the wires through the rotor arms—you'll want about 150mm of cable to play with at either end of the arms.

- Insert four 50mm M3 machine screws through the underside of the PDB (power distribution board) and secure on the topside with lock nuts. (These will form part of your standoff for the flight controller.) Attach the PDB where you drilled the 25mm hole, with the battery connectors facing the hinge of the lid. I used sticky velcro tape because I knew I might need to take it off. As you are attaching the power board, pull through the wires three at a time and attach a small cable tie.
- You'll have great fun now squeezing all this in and making it look tidy and safe! It's worth adding some hot glue to the PDB connections as they are not very well insulated. Once you're confident it's all connected correctly, it's time to install the flight controller.

Step #5: Install the Flight Controller

- Cut out a scrap of plastic that will fit inside your box and support your flight controller (FC). Drill four holes in the plastic that will align with the machine screw standoffs on your PDB.

- Attach your ESC control wires to the FC. Slide some 10mm round spacers on the machine screws on the PDB. Fix the FC to the plastic support (I used double-sided tape as indicated by DJI) and gently push it down onto the spacers, then attach lock nuts and screw them tight.

Step #6: Prepare the Batteries

- It's very important to get the specified battery size that will fit under the lid of the box. Each battery's dimensions are: 102mm × 37mm × 24mm.

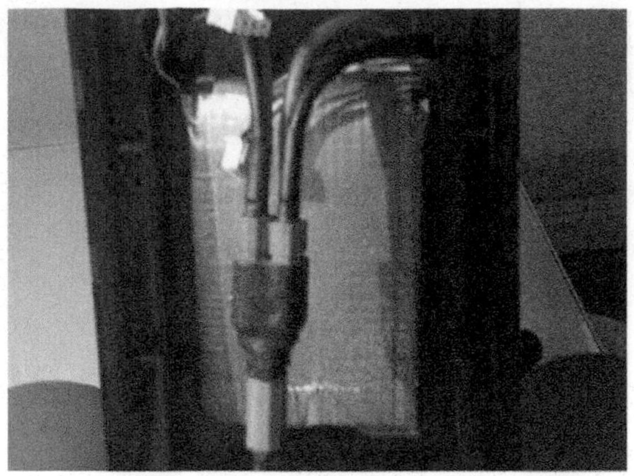

- I just used gaffer's tape to bind the two batteries together and then used a XT60 harness battery splitter to wire them into the PDB.

Step #7: Install the GPS Puck

- Drill a 50mm hole in the center of the hinged lid. Then glue the 63.5mm plastic dished head on the top of the lid with superglue or PVC pipe weld. Key the surface lightly with sandpaper to help the glue.

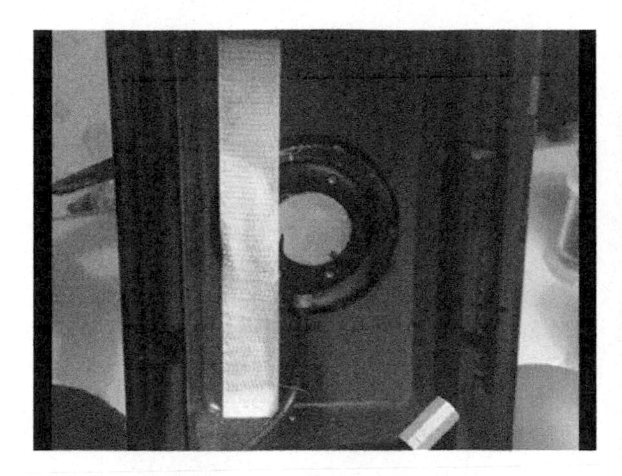

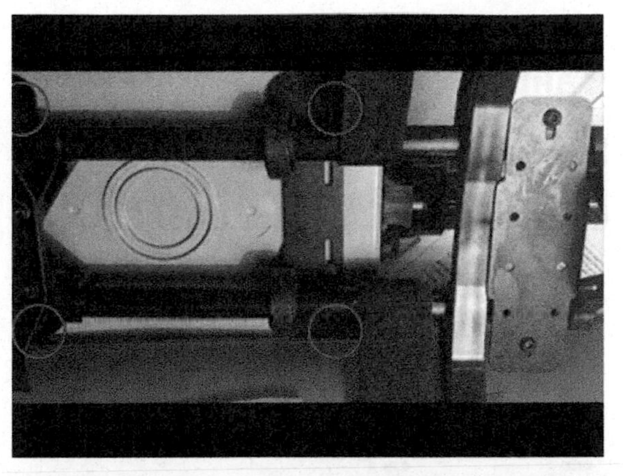

- Install the GPS puck in the hole in the lid and secure it with strips of velcro tape. These will do double duty to secure the battery pack as well.

Step #8: Install Undercarriage and Floats

- Prepare your undercarriage. The landing gear kit I bought had two trays that attach to the frame, where you can easily attach camera accessories, FPV kit, etc.

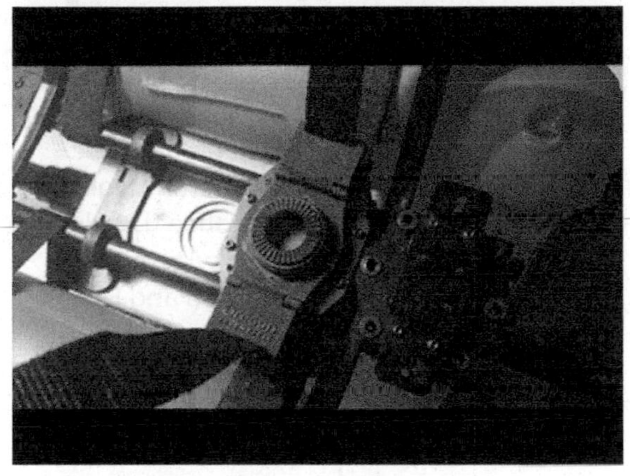

- Now attach the floats. I used velcro straps to attach them to the undercarriage. Your installation may vary.

Step #9: Flight Check and Calibration

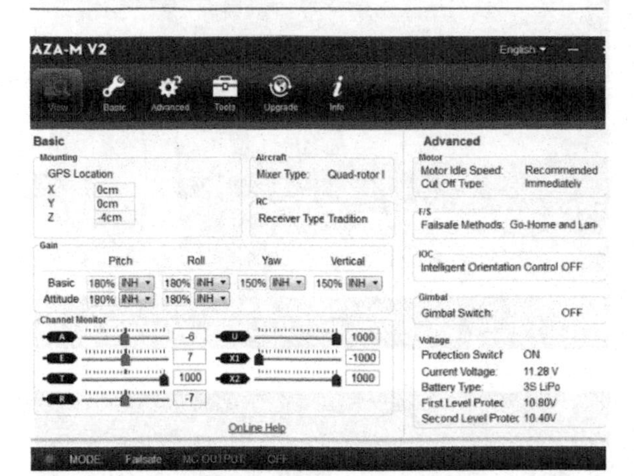

- I won't go into great detail here because there are many different flight controllers. Also, you'll notice I only have a six-channel TX/RX system and this isn't ideal when using GPS and RTH functions on the Naza. I'll be upgrading the drone to an ArduCopter flight system in the near future and let you know how I get on.
- If you're using the DJI Naza-M Lite setup, you'll find instructions online for calibrating the controller with your computer via USB: *http://bit.ly/21KzqYT*.

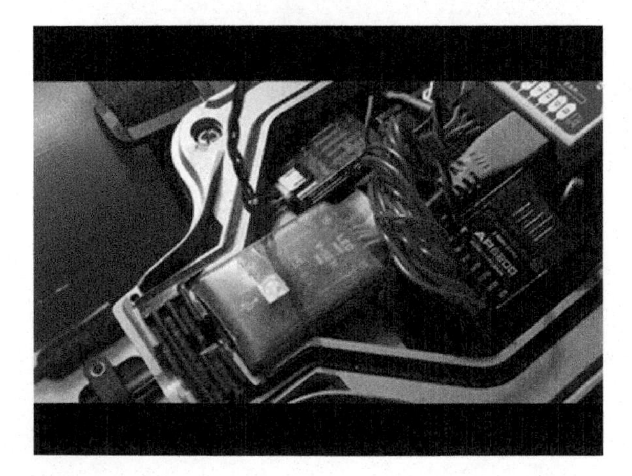

- One nice thing about the Naza is how perfectly snug the VU fits under the clip of the main box.
- Install your motors, without props. As long as they're brushless, which most are, then no waterproofing is necessary. It's worthwhile spraying them after water takeoffs and landings with silicone spray to dry them out. (I've also heard great things about Liquipel; if you try it, let us know how it works for you.

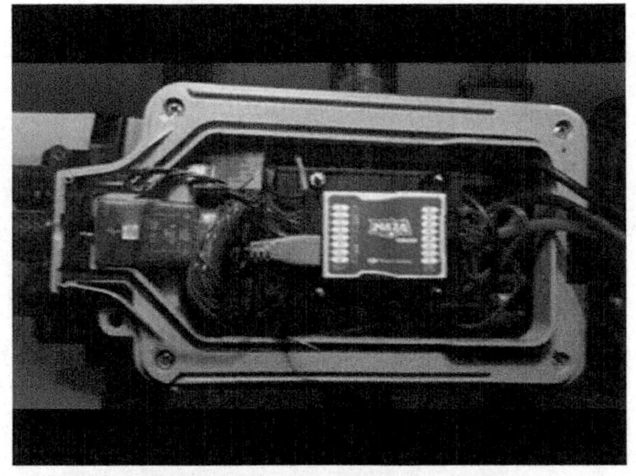

- Connect the batteries and test all your flight control systems on the drone before attaching the props to the motors.

Step #10: Final Checks and Sealing

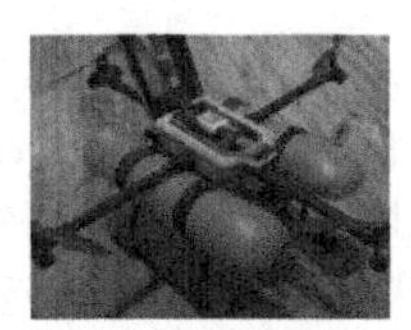

- It's now time to double-check that all seals are good. If any of the motor mounts or reducers are twisting in the frame, detach them and add a small amount of PVC weld to secure them.

- Don't add any PVC weld to the round 4-way junction box. If (and when) an arm breaks, you'll easily be able to remove and replace that one arm rather than remove the entire hub.
- Happy flying!

Payload Box and Drop Mechanism for Drones and R/C Planes

13

When Uplift Aeronautics founder Mark Jacobsen envisioned using a fleet of drones to drop food and medical supplies on war-torn regions, he needed a device to make that happen. The answer was to modify fixed-wing foam UAVs, and the result was five different designs, including the Waliid, pictured here.

—*By Michael Thomas Taylor*

Figure 13-1 *Photo by Hep Svadja (all other photos courtesy of Uplift Aeronautics)*

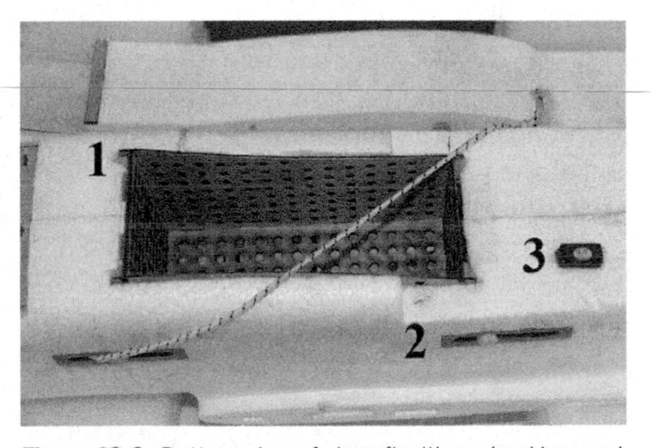

Figure 13-2 *Bottom view of aircraft with payload box and drop mechanism installed*

This project is a set of instructions for building and installing a payload box and drop mechanism for the X-UAV Talon R/C airplane. Design credit goes to Brandon Fetroe of Uplift Aeronautics.

The files for the custom parts can be found on Uplift's Github at *https://github.com/upliftaero/waliid*.

Parts/Tools

- Wood, 1/8″ sheets
- Nylon bolts, four 1″, four 1.25″, 20 threads per inch (8)
- Nylon nuts, 20 threads per inch (8)
- Metal washer
- Servo, 9g (micro size) analog metal
- Metal pin/control rod
- Metal tube, 1/2″ long, 1/4″ diameter guide hole for pin

- Bungee cord
- X-UAV Talon kit

The system consists of three major components:

- The payload box provides a space to store and secure the payload in the fuselage of the X-UAV Talon. It is recessed into the body of the aircraft through a hole cut in the bottom of the fuselage.
- The payload box is secured to four mounting brackets glued into the sides of the airframe. Four nylon bolts attach from the outside to hold it in place.
- The drop mechanism is a servo that extends and retracts a metal pin poking through the side of the body. A bungee cord wraps around the bottom of the payload and attaches to the pin to hold it in place.

Figure 13-3 *The bungee secures to the pin of the drop mechanism; the servo retracts the pin to release the bungee and drop the payload*

The package is secured inside the airplane by a bungee cord tied at one end to the wooden plate in the nose of the aircraft. It is then stretched across the payload box opening to a pin on the opposite side. When the servo retracts the pin, the bungee will release and drop the payload.

Most of the custom parts were laser cut out of 1/8″ or 1/16″ plywood, with files provided in the Git repository. In the absence of access to a laser cutter, most parts can be cut with hand tools, omitting the grid of holes in the box (done to save weight). SolidWorks part drawings are also available to work from in the repository.

Step #1: Construct Payload Box

1. The payload box provides a space to hold the payload inside the aircraft. It is designed to hold a single 3"×7.5"×6″ cardboard box.
2. Assemble the front, back, top and two side plates to make an open-sided box as shown. Secure the parts with wood glue or superglue. The taller side will face towards the front of the aircraft, and will be referred to as the "front" of the box.

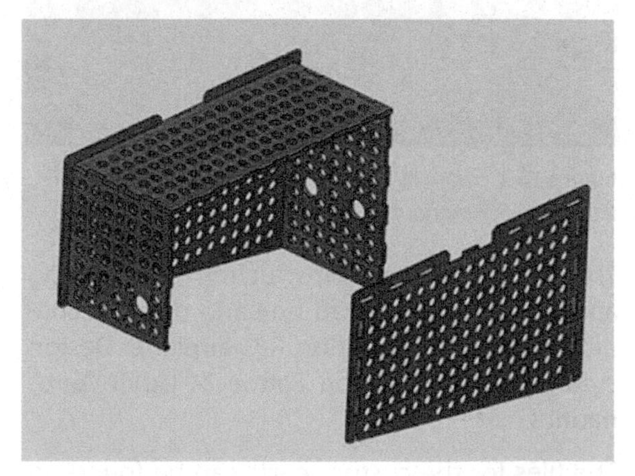

3. Attach the bolt plate to the front and rear of the box. The plate must be aligned with the large holes in the front facing side. The plate serves as a mounting point for a pair of 1/4″ 20 nylon bolts that will secure the payload box to the mounting bracket.

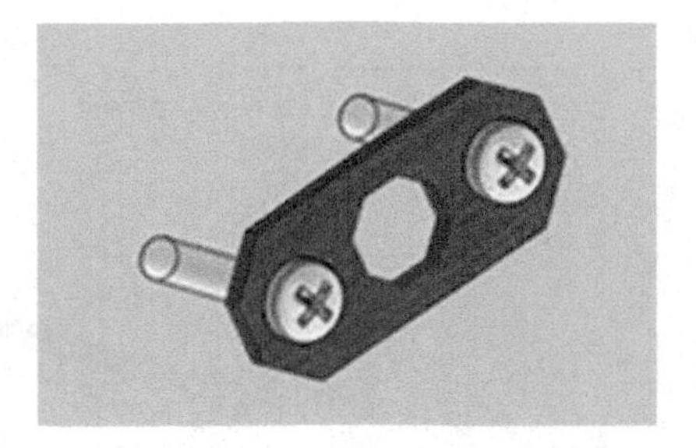

4. Align the plate so the bolts slot through perpendicular to the box as shown. DO NOT GLUE THE BOLTS TO THE PLATE—the bolts are meant to be inserted after the box has been placed in the airframe.
5. Attach the second bolt plate to the rear of the payload box.

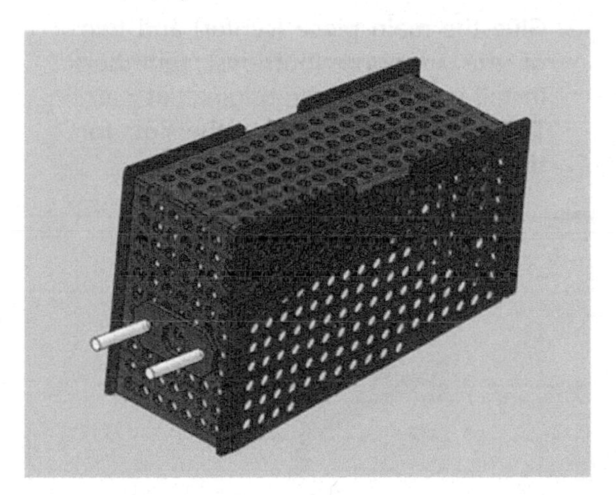

6. (Optional) Strengthen the corners of the payload box with segments of balsa wood.

Step #2: Build Mounting Bolt Assembly

1. The payload box is held inside the aircraft by four nylon bolts that attach from the outside. To give these bolts somewhere to attach to, a mounting assembly attaches to the end of the box.

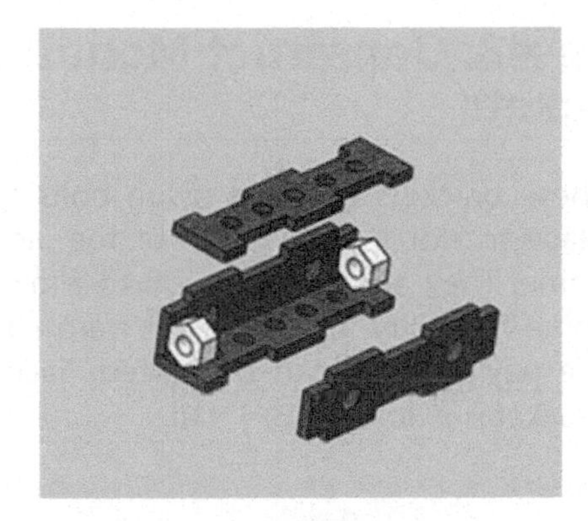

2. The payload box (1) is attached to the mounting assembly (2), which is attached to the airframe by the bolts (3).

3. Each assembly has two nylon nuts that will hold the external bolts, securing the whole payload assembly to the airframe. Use superglue to assemble three sides of the wooden parts, then glue the nylon nuts inside each end. Then glue the last wooden piece to complete the assembly.

Step #3: Construct Mounting Brackets

1. These brackets give the nylon bolts a place to slot into the side of the airframe. They are long enough to allow some flexibility in how far forward/aft the payload box can be placed. Each bracket is 3″ long by 3/4″ tall.

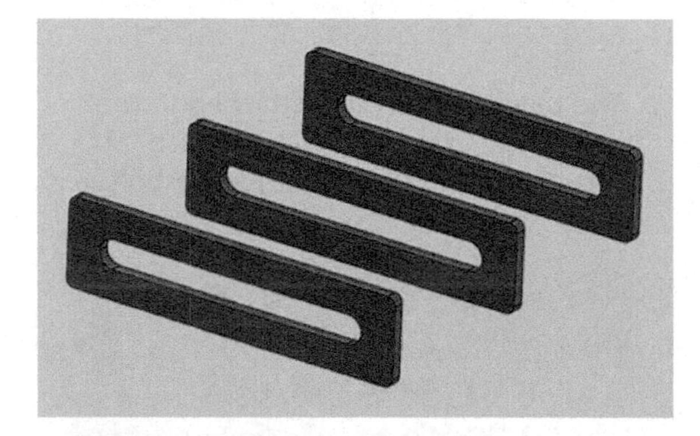

2. Assemble the mounting brackets by stacking three of the laser cut pieces as shown and gluing them together.

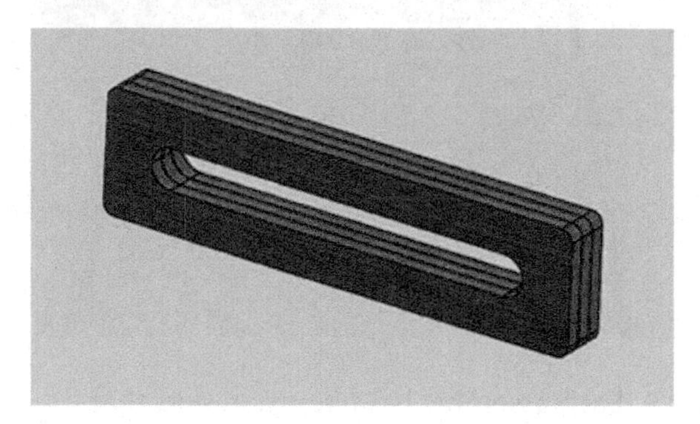

Step #4: Assemble Front Brace

1. The final custom part is a new front brace for the wooden frame in the center of the fuselage. The new front brace replaces the stock front brace in order to provide enough room to fully insert the payload box into the airplane.

2. Glue the main piece (center) and two of the supports (outsides) together. Install in place of the original kit part. The middle piece fits into the slots for the stock kit front brace.

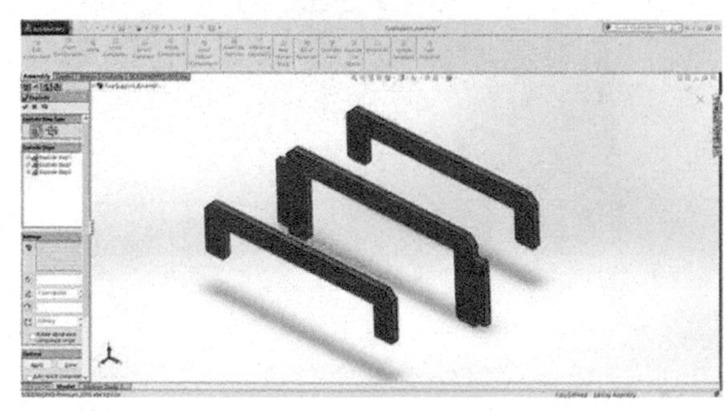

Step #5: Modify Airframe

1. Cut the payload box hole in the bottom of the airframe. The hole should be along the centerline. The 5.5″ from the rear lip is approximate. The hole can be moved more fore/aft to shift the center of gravity. It may be necessary to cut notches for the bolt plate and edges of the box to fit.

2. Then cut the mounting bracket holes. It is important that the mounting brackets are in line with the mounting assembly when the payload box is installed.

3. It is a good idea to install the payload box and take note of the mounting assembly locations before cutting the holes for the mounting brackets. (See Steps 1 and 2.)

4. Cut two 3"× 3/4" rectangular holes in the side of the fuselage to fit the mounting brackets. Adjust the location of the holes fore/aft depending on where the box will finally sit in the airframe.

5. If you are modifying a new airframe, the top of the holes for the mounting braces align with the interior foam molding as shown.

Step #6: Install Front Brace

1. Install the new front brace for the wooden frame. If the kit has not been assembled yet, simply replace the old front brace with the new part when building the wooden frame for the center of the fuselage. If modifying an already built kit, cut the old X brace and glue on the outside pieces from Step 4 to strengthen it.

Step #7: Install Bungee and Payload Box

1. Tie a length of bungee cord to the front plate in the nose. It should be long

enough to reach out one front mounting bracket and across the hole cut for the payload to the other side rear mounting bracket (see photo in intro). Feed the other end through one of the front mounting brackets and tie a metal washer on the end.

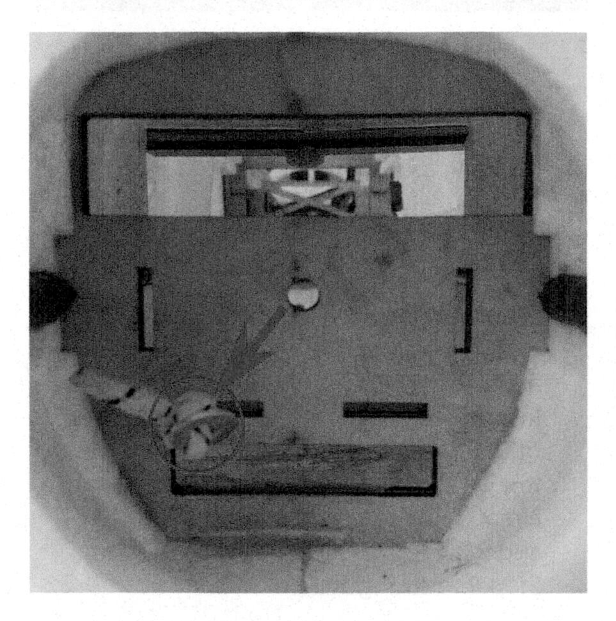

2. Insert the payload box into the hole cut in the bottom of the fuselage. The tall end faces the front of the aircraft. It should be nearly completely recessed into the airframe.

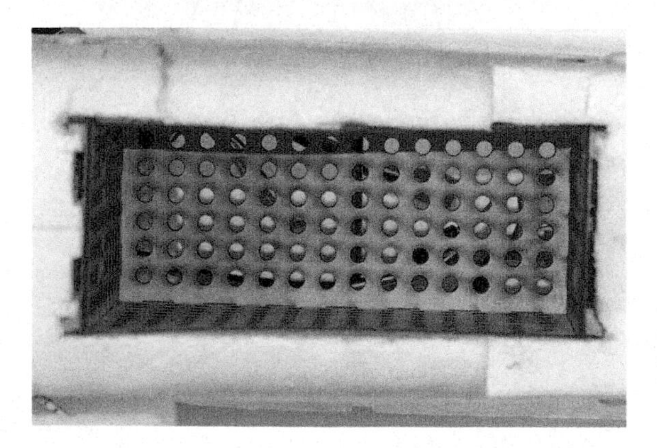

Step #8: Bolt It In

1. Place two 1″ long 1/4″ nylon bolts through the holes in the bolt plate on the front and back of the box once it is in the airframe.

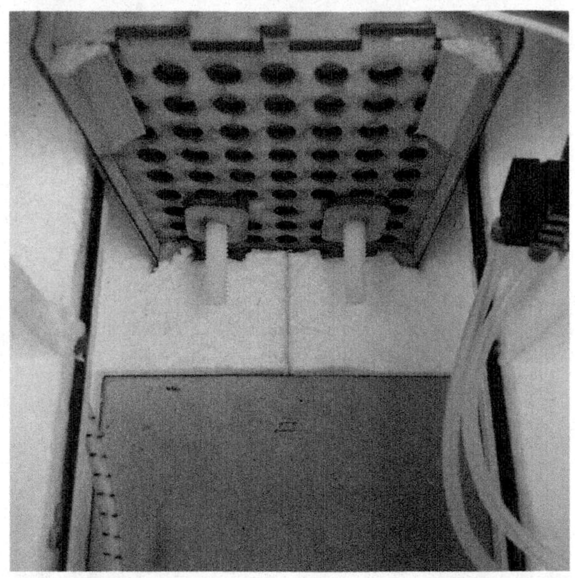

2. Note: The bolt plates in the photo are from an older design; see Step 1 for an updated design.

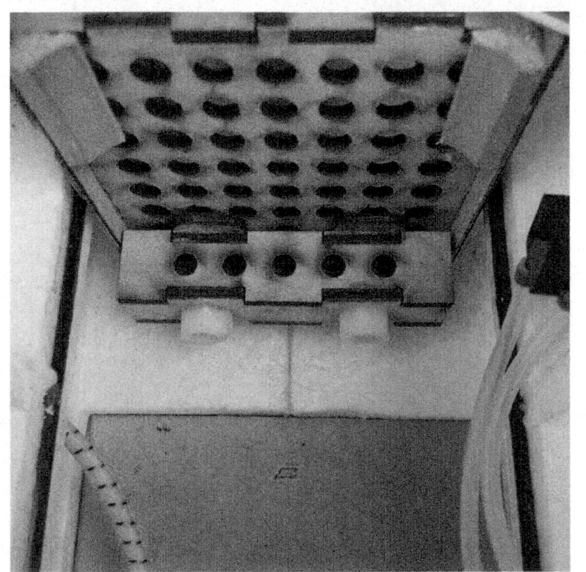

3. Place the mounting assembly over the bolts on the box. Secure with nuts. Secure the payload box with 1.5" long nylon bolts through the mounting brackets into the nuts (as seen in Step 2) in the mounting assembly. 4X: Front and back, and on both sides.

Build Your First Tricopter

14

They fly smoother and make better videos than quads. Build the Maker Hangar Tricopter and see for yourself!

—From Make:44 by Lucas Weakley

Quadcopters are a little easier to build, but tricopters have advantages that make them more exciting to fly—especially for shooting aerial video. I built my first one in 2010, inspired by David Windestal's beautiful aerial GoPro videos. I didn't get many flights out of that first build, but I learned a lot. After building several more, I've developed an affordable kit that anyone can build—the Maker Hangar Tricopter.

Why Fly Tri?

A tricopter's three motors are usually separated by 120°, not 90° like a quadcopter's. This makes them great for video because you can place the camera really close to the body and still have no propellers in view. And where quads must rely on counter-rotating propellers to handle torque and balance the aircraft, a tricopter can use identical props because it has a special servo in the back—a yaw servo—that twists the tail motor to counter torque.

Tricopters fly differently, too. With their dedicated motor for yaw (turning), they fly with more fluid, natural-looking movements—they can bank, pitch, and yaw like an airplane, but still hover like a helicopter. A quadcopter's flight is more robotic, as the controller board calculates the precise rotation for all four motors to create the proper torque and balance to yaw the aircraft. If you let go of the stick, a quad stops turning abruptly; for video work, this can be obvious and distracting. Let go of a tricopter's stick and the tilted tail motor takes a moment to return to a hovering position; this gives you a slow stop and even a little overshoot, as though a person were moving the camera.

Finally, tricopters are a lot of fun to fly, especially for stunts and acrobatics. The tilting motor also gives you much higher yaw speeds —that means they turn faster.

A Tricopter for Makers

The Maker Hangar Tricopter is made of wood—hackable, easy to drill and cut, and a natural absorber of vibration, the enemy of aerial video. The airframe is big, with plenty of room for large controller boards, video transmitters, drop mechanisms, or whatever you can imagine. And we widened the front arms to about 150° so our tricopter is more agile.

The kit includes a 3D-printed tail assembly and all the hardware you'll need, plus a wire rope vibration absorber that will pretty much erase camera vibrations even if your propellers are unbalanced. A carbon-fiber hinge provides a strong, smooth connection between the tail motor and airframe.

Finally, like most tricopters, the two front arms lock in place for flight, then fold back neatly for transportation and storage.

Parts

- Maker Hangar Tricopter Kit—$85 from *http://bit.ly/1D9wNWU*

The kit includes:

- Laser-cut plywood airframe parts (download the files from Dropbox)
- 3D-printed tail assembly (download the files from Dropbox)
- Carbon-fiber hinge pieces
- Oak square dowels, 7/16"×7/16"×12" (3) for the arms
- Bolts, stainless steel, M3: 25mm (8), 6mm (4), 10mm (16), and 22mm (8)
- Lock nuts, M3 (25)
- Washers, M3 (16) and M4 (2)
- Bolts, nylon, 6-32×3/8" (4)
- Nuts, nylon 6-32 (4)
- Standoffs, 6-32×1-1/2" (4)
- Cable ties (20)
- Push rods, 2-1/2"×0.047" (2)
- Push rod connectors (2)
- Velcro straps (2)
- Wire rope, 3" lengths (4)

Electronics (not included)—see the kit web page for complete recommendations:

- Flight controller board (see Step 12 below)
- R/C receiver to match your R/C transmitter
- Motors, brushless outrunner, 900kV (3) Emax GT2215/12
- ESCs, 20A (3) Emax Simon
- Props, 10×4.7 (3)
- Batteries, LiPo, 3,300mAh (2)
- Servo, micro
- Servo extension, 6"
- Wire, 16 gauge stranded
- Heat-shrink tubing
- Servo cable, male to male
- JST connector (optional)

Tools

- Drill and bits
- Pliers, needlenose
- Pliers, side cutting
- Wire cutters/strippers
- Hot glue gun
- Cyanoacrylate (CA) glue
- Screwdriver
- Hex driver set
- Adjustable wrench
- Sandpaper
- File
- Hobby knife
- Soldering iron and solder
- Heat gun or hair dryer
- Helping hands (optional)

It's a great kit for anyone wanting to get into multicopters or aerial photography. You can also build it totally from scratch: download the PDF plans, laser cutter layouts, 3D files for printing, flight controller settings, and watch the how-to video series in this Dropbox folder (*http://bit.ly/25iBADF*).

Specs

- Flight time: 12 minutes
- Frame weight: 325g
- Flight weight: 1kg
- Compatible with 8″–10″ props
- Wire rope vibration absorber
- 22mm motor mounts

Step #1: Sand and Paint

Sand down any burrs or splinters on the wooden parts. If you wish, paint with a couple of light coats.

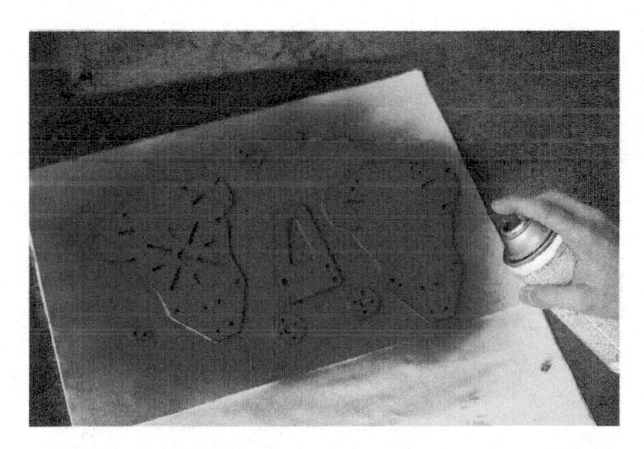

Step #2: Assemble the Hinged Tail

To build the hinge, glue the 2-1/2″ carbon rod flush into the 3/4″ carbon tube using CA glue. Hot-glue this end into the 3D-printed motor mount. Also hot-glue the 1″ carbon tube into the 3D-printed tail piece.

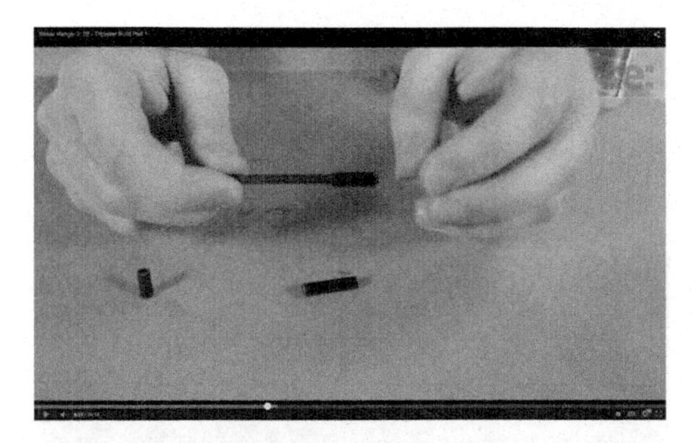

Now put it together: slide an M4 washer on the hinge rod, then the tail piece, then another washer. Finally, glue the 1/2″ carbon tube to the end of the rod to capture the whole assembly.

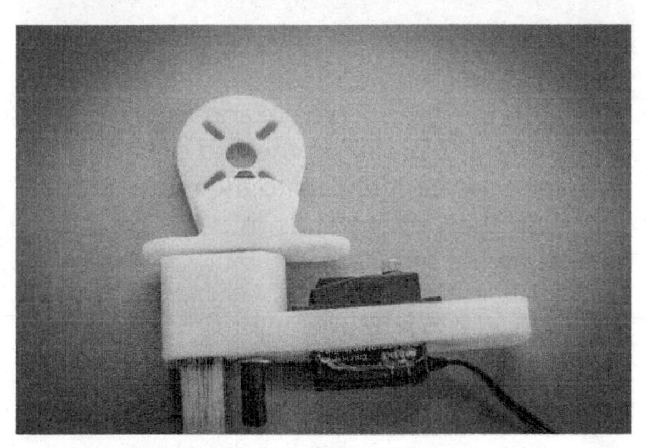

Hot-glue the servo into the tail piece and install two "easy connectors" in 1/16″ holes on the servo arm. You can glue the hardwood tail arm into the tail piece now as well.

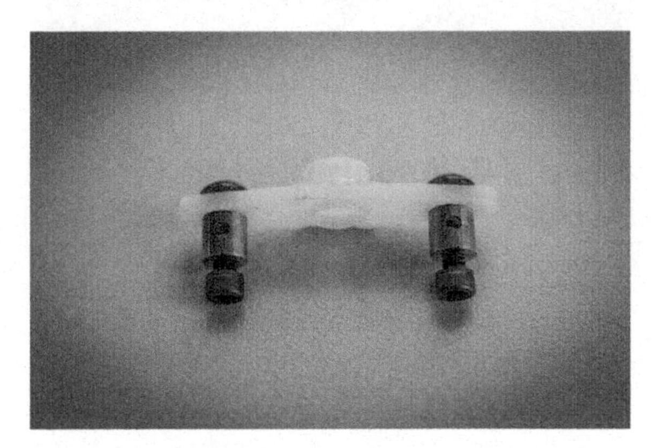

Step #3: Assemble the Hinged Tail (Cont'd)

Bolt the tail motor into the motor mount with M3 washers.

Finally, connect the servo linkages. Use pliers to create a tiny "Z-bend" on the end of each push rod. Hook the bent ends into the motor mount, and slide the unbent ends into the easy connectors on the servo arm.

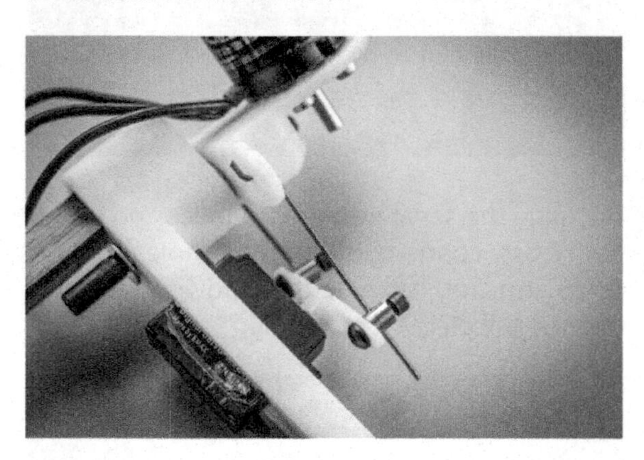

Step #4: Assemble the Front Arms

Drill each front arm using the two templates provided: at one end for the motor mounts, and at the other end for the rotation bolts for folding the copter arms.

Note that the motor template has two different spacing patterns; use each pattern on only one arm, so your motors will end up being mounted symmetrically.

Step #5: Assemble the Front Arms (Cont'd)

Then mount the remaining two motors using the four round plywood motor mount pieces—the ones with the larger center holes go up against the motors—and M3×22mm bolts with washers.

Step #6: Prepare the Body Plates

Install the nylon standoffs on the upper body plate, but don't mount the small top plate yet.

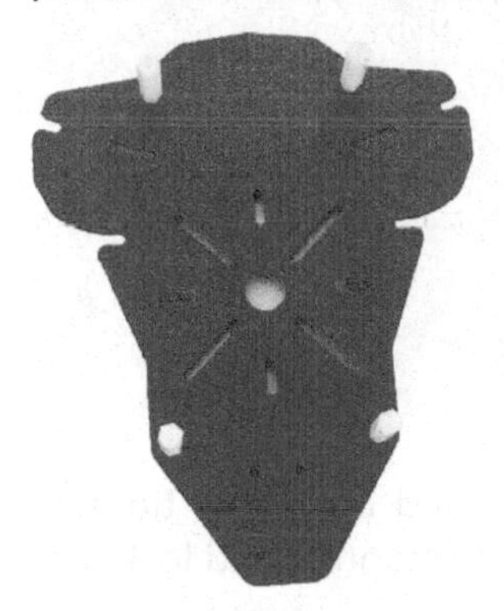

Bolt four of the small plywood brackets to the lower body plate, and four to the camera/battery tray, using M3×10mm bolts and nuts.

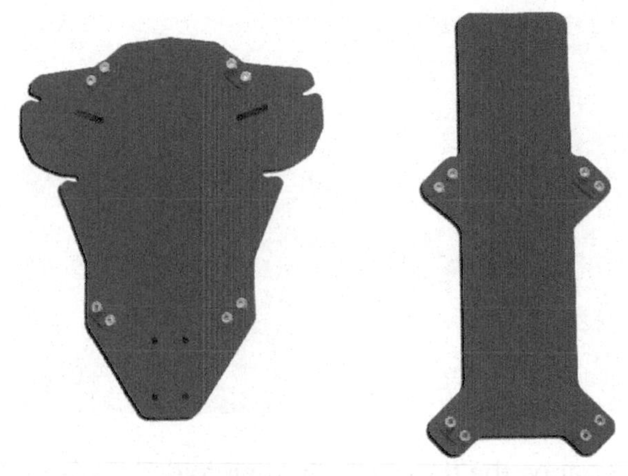

This tray is optional (you could just velcro the battery to the bottom of the copter) but it's highly recommended for video because it's isolated from vibrations by short wire ropes. Clamp the four wire ropes into the brackets on the bottom plate, but don't connect the camera tray yet.

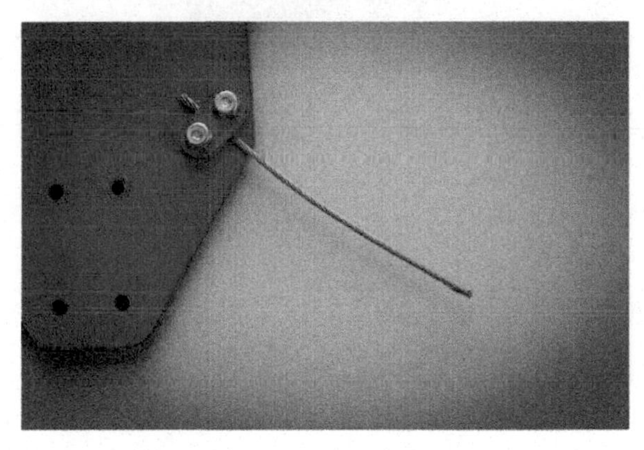

Step #7: Install the ESCs

Connect the three electronic speed controllers (ESCs) to the motors and zip-tie them to the arms.

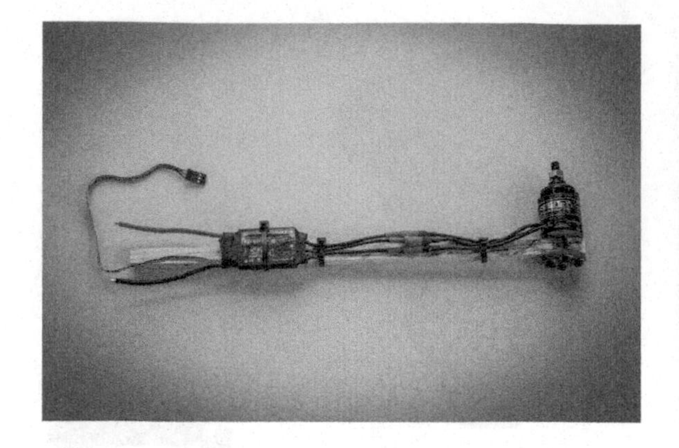

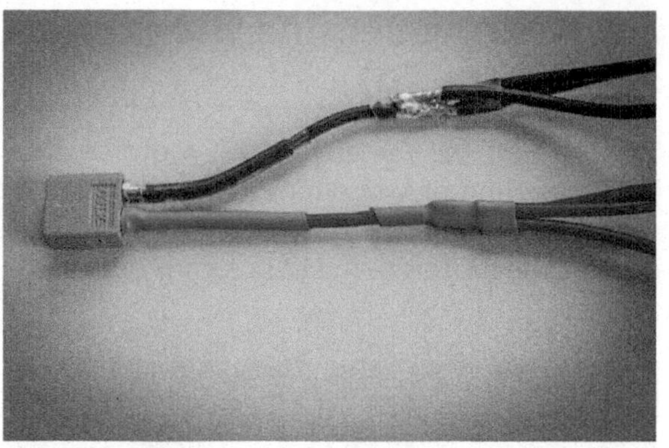

Arrange the three arms in their folded configuration, then measure out enough wire to extend all the power and ground wires to meet at the back of the body. Solder the extension wires and insulate connections with heat-shrink tubing. Strip the free ends and solder them into your battery connector.

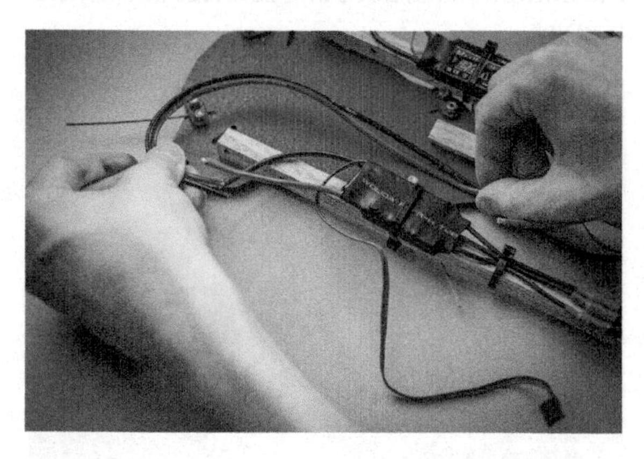

Now's also the time to splice in a JST connector (optional) if you want to power an onboard FPV (first-person video) system and watch live video from the tricopter. Learn more about batteries, FPV, and other flight components in the first season of Maker Hangar videos.

Step #8: Attach the Arms

Bolt the two front arms to the lower body plate through the outer mounting holes, using M3×25mm bolts and lock nuts. Place the upper body plate on top, then pass two more bolts through the locking slots and the inner arm holes, and secure with washers and lock nuts. Finally, clamp the tail arm between the body plates using four bolts.

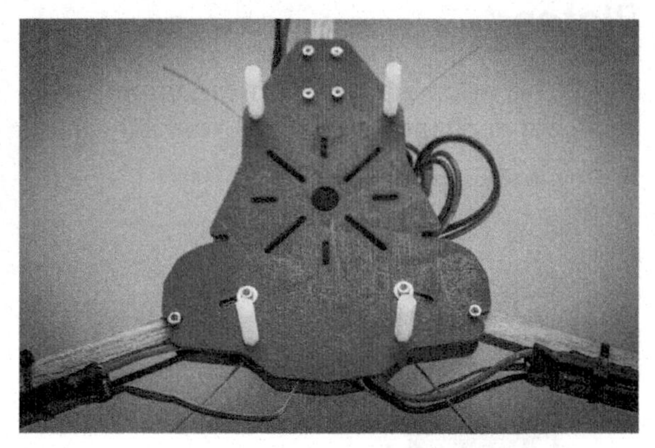

Test the folding action and loosen or tighten bolts until the arms fold smoothly and lock forward securely.

Step #9: Mount the Landing Gear

Zip-tie the two plywood landing struts to the front arms.

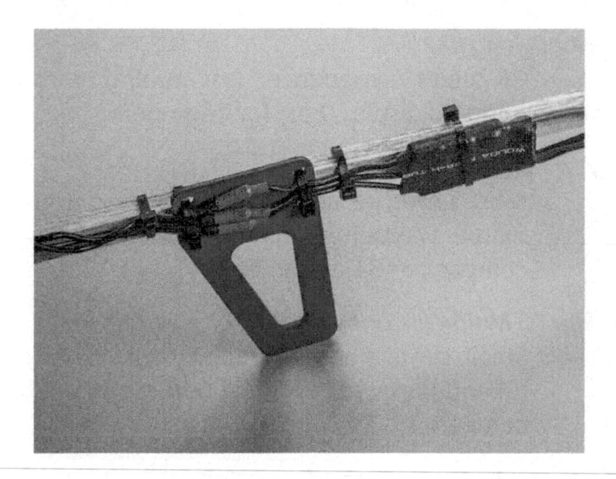

Step #10: Suspend the Camera Tray

Clamp the free ends of the wire ropes into the brackets on the camera tray. Make sure the camera platform faces forward and the bolt heads face outward; you'll need access to them to adjust the tray later. Strap the battery to the tray with the velcro strap.

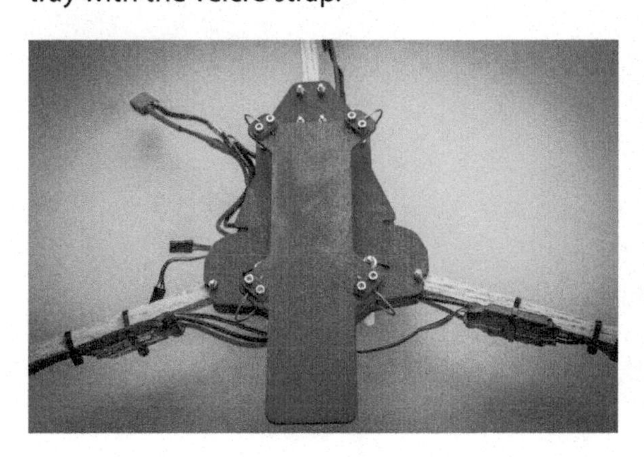

Step #11: Mount the Receiver and Flight Controller

Attach your flight controller to the upper body plate using hot glue, double-sided tape, or bolts through the mounting slots. (We used the Flip 1.5 MWC controller. You can download the settings at the Maker Hangar project page.)

Bind your R/C receiver to your transmitter (see Maker Hangar Season One, Episode 12), and then set the throttle ranges by plugging each of your ESCs, in turn, into the receiver's Throttle port (Season 2, Episode 4). Mount the receiver and plug it into the flight controller. Center the yaw servo and tighten the linkages.

Finally, screw the top plate to the standoffs to protect your electronics, and your Maker Hangar Tricopter is complete!

Step #12: About Flight Controllers

The flight controller board converts the signals from your transmitter into the motor speeds that move your tricopter. It also reads the aircraft's position and movements with its onboard gyros and accelerometers, and makes tiny changes to motor speeds to counter the wind, torque, and other forces that are trying to tip the copter over.

These are the boards I recommend for the Maker Hangar Tricopter:

OpenPilot CC3D
: The best flight experience, easy setup, but tuning takes time

HobbyKing KK2
: OK flight experience, fast tuning with onboard display, best for beginners

ArduPilot APM 2.6
: Most powerful and expensive; programmable waypoint capabilities with GPS, compass, and barometer

Flip 1.5 Multi Wii Controller (MWC)
: Small, simple, and affordable, but powerful and flies well; optional barometer and compass

Index

Symbols

A

B

C

D

E

F

G

H

Colophon

The cover photo is by Mark Harrison. The cover and heading font is Benton Sans; the body font is Myriad Pro.